FLORIDA STATE
UNIVERSITY LIBRARIES

OCT 17 1995

TALLAHASSEE, FLORIDA

Analyzing Superfund

Economics,
Science,
and Law

Analyzing SUPERFUND

Economics, Science, and Law

Edited by
Richard L. Revesz and
Richard B. Stewart

Resources for the Future
Washington, DC

©1995 Resources for the Future

All rights reserved. No part of this publication may be reproduced by any means, either electronic or mechanical, without permission in writing from the publisher, except under the conditions given in the following paragraph.

Authorization to photocopy items for internal or personal use, the internal or personal use of specific clients, and for educational classroom use is granted by Resources for the Future, provided that the appropriate fee is paid directly to Copyright Clearance Center, 222 Rosewood Drive, Danvers, MA 01923, USA.

Printed in the United States of America

Published by Resources for the Future
1616 P Street, NW, Washington, DC 20036-1400

Library of Congress Cataloging-in-Publication Data

Analyzing superfund: economics, science, and law/ Richard L. Revesz and
    Richard B. Stewart, editors.
        p.   cm.
    Includes bibliographical references and index.
    ISBN 0–915707–75–6

       1. Pollution—Economic aspects—United States. 2. Transaction costs.
3. Liability for hazardous substances pollution damages—United States.
4. Hazardous substances—Law and legislation—United States.
5. Hazardous waste site remediation—United States. 6. Environmental
remediation—United States. 7. United States. Comprehensive
Environmental Response, Compensation, and Liability Act of 1980.
I. Revesz, Richard L., 1958– . II. Stewart, Richard B.
HC110.P55A59 1995
363.72'87'0973—dc20   94-41681
                                                                                   CIP

∞ The paper in this book meets the guidelines for permanence and durability of the Committee on Production Guidelines for Book Longevity of the Council on Library Resources.

This book is the product of the Center for Risk Management at Resources for the Future, Terry Davies, director. It was copyedited by Eric Wurzbacher. The book and its cover were designed by Diane Kelly, Kelly Design.

RESOURCES FOR THE FUTURE

Directors

Darius W. Gaskins Jr., *Chair*
Paul C. Pritchard, *Vice Chair*
Anthony S. Earl
Lawrence E. Fouraker
Robert W. Fri
Robert H. Haveman
Donald M. Kerr
Frederic D. Krupp

Henry R. Linden
Thomas E. Lovejoy
Karl-Göran Mäler
Frank D. Press
Robert M. Solow
Linda C. Taliaferro
Victoria J. Tschinkel
Mason Willrich

Officers

Robert W. Fri, *President*
Paul R. Portney, *Vice President*
Edward F. Hand, *Vice President–Finance and Administration*

RESOURCES FOR THE FUTURE (RFF) is an independent nonprofit organization engaged in research and public education on natural resource and environmental issues. Its mission is to create and disseminate knowledge that helps people make better decisions about the conservation and use of their natural resources and the environment. RFF neither lobbies nor takes positions on current policy issues.

Because the work of RFF focuses on how people make use of scarce resources, its primary research discipline is economics. Supplementary research disciplines include ecology, engineering, operations research, and geography, as well as many other social sciences. Staff members pursue a wide variety of interests, including the environmental effects of transportation, environmental protection and economic development, Superfund, forest economics, recycling, environmental equity, the costs and benefits of pollution control, energy, law and economics, and quantitative risk assessment.

Acting on the conviction that good research and policy analysis must be put into service to be truly useful, RFF communicates its findings to government and industry officials, public interest advocacy groups, nonprofit organizations, academic researchers, and the press. It produces a range of publications and sponsors conferences, seminars, workshops, and briefings. Staff members write articles for journals, magazines, and newspapers; provide expert testimony; and serve on public and private advisory committees. The views they express are in all cases their own and do not represent positions held by RFF, its officers, or trustees.

Established in 1952, RFF derives its operating budget in approximately equal amounts from three sources: investment income from a reserve fund; government grants; and contributions from corporations, foundations, and individuals. (Corporate support cannot be earmarked for specific research projects.) Some 45 percent of RFF's total funding is unrestricted, which provides crucial support for its foundational research and outreach and educational operations. RFF is a publicly funded organization under Section 501(c)(3) of the Internal Revenue Code, and all contributions to its work are tax deductible.

Contents

Foreword   ix
Paul R. Portney

Preface   xi
Richard L. Revesz and Richard B. Stewart

PART I: Introduction

1. **The Superfund Debate**   3
   Richard L. Revesz and Richard B. Stewart

PART II: The Cleanup Standards

2. **Confronting Superfund Mythology: The Case of Risk Assessment and Management**   25
   Katherine D. Walker, March Sadowitz, and John D. Graham

3. **The Magnitude and Policy Implications of Health Risks from Hazardous Waste Sites**   55
   James T. Hamilton and W. Kip Viscusi

4. **Do Benefits and Costs Matter in Environmental Regulation? An Analysis of EPA Decisions under Superfund**   83
   Shreekant Gupta, George Van Houtven, and Maureen L. Cropper

PART III: The Liability Regime

5. **Evaluating the Effects of Alternative Superfund Liability Rules**   115
   Lewis A. Kornhauser and Richard L. Revesz

6. **Evaluating the Impact of Alternative Superfund Financing Schemes**   145
   Katherine N. Probst

PART IV: Transaction Costs

7. The Transaction Costs Generated by Superfund's
   Liability Approach — 171
   Lloyd S. Dixon

8. De Minimis Settlements under Superfund: An Empirical
   Study — 187
   Lewis A. Kornhauser and Richard L. Revesz

PART V: Natural Resource Damages

9. Liability for Natural Resource Injury: Beyond Tort — 219
   Richard B. Stewart

APPENDIX

Conference Agenda and Participants — 251

INDEX — 257

Foreword

The Comprehensive Environmental Response, Compensation, and Liability Act of 1980, better known as Superfund, has grown to be one of the most controversial of all environmental laws. The 103rd Congress came very close to amending the law in 1994, and the changes contemplated would have addressed several of its most vexing features. Despite this legislative "near-miss," however, the debate over proposed changes was generally uninformed by serious policy analysis. In its place, one anecdote tended to be piled atop another.

In *Analyzing Superfund: Economics, Science, and Law,* New York University law professors Richard Revesz and Richard Stewart have assembled a volume that should help enlighten the Superfund debate when it begins again in earnest in 1995. The chapters in this book, written initially as papers for a conference held at New York University in the fall of 1993, address issues that are central to the reauthorization debate and that will endure long after the legislative action is completed. For example, the chapters examine the hotly contested liability standards written into the law in 1980, as well as the transaction costs to which this liability gives rise. They also provide information and analysis relevant both to the selection of cleanup approaches at Superfund sites and to the determination of natural resource damages for which responsible parties may be liable even after sites have been cleaned up.

As the title indicates, this book addresses Superfund from a variety of approaches. Several of the chapters deal almost exclusively with the results of quantitative risk assessments conducted at Superfund sites. Although these chapters draw out the policy implications of the risk estimates, they are grounded in the science of risk assessment. Two other chapters are squarely in the realm of law and economics. They shed light on the effects that alternative liability rules might have on cleanup actions, and on the problems that might arise under the current provisions for assessing damages to natural resources under Superfund. Several other chapters present carefully developed estimates of the economic consequences of Superfund, either in terms of the annual costs of the program as a whole, the expenditures on legal fees and other transaction costs, or the pattern of settlements in cases where the government has sued private parties over cleanup liability. There should be something in this volume to tempt all manner of readers.

If Congress waits long enough to amend the Superfund law, a satisfactory factual base may actually exist upon which which Congress can base its deliberations. This book edited by Revesz and Stewart is a welcome addition to the small but rapidly growing collection of analytical studies of Superfund. It will be interesting reading for both policymakers and academics who must struggle to make sense of and help improve environmental regulation in the United States.

Paul R. Portney
Vice President
Resources for the Future

Preface

This book grows out of the Conference on Superfund Reauthorization: Theoretical and Empirical Issues, which took place at New York University School of Law on December 3–4, 1993. Eight of the nine chapters in the book are the principal papers that we commissioned for the conference. In addition, we wrote an introductory chapter that provides a background to the Superfund statute and explains how the various contributions fit into the debate over the future of Superfund. The discussions at the conference and the final versions of this book's chapters were greatly enhanced by a distinguished group of policymakers at the federal and state levels, academics, and representatives of the various groups with a stake in Superfund who served as the commentators and moderators of the various panels. Their names and affiliations are set forth in the Appendix.

We convened the conference at a time when the political debate over the reauthorization of Superfund was getting underway. Our immediate hope was to inform, however modestly, the highly charged political debate on the fate of Superfund. The failure by the 103rd Congress to reauthorize Superfund in 1994 means that the debate will be renewed in the new political context of the 104th Congress. We hope that the book will assist the renewed analysis and evaluation of the current program. More ambitiously, we also hope for a long-term payoff—to illuminate the basic issues involved in the design of programs of environmental liability and remediation. We believe that the chapters in this book meet both objectives.

We owe many debts of gratitude. The conference was funded by a grant from Freeport-McMoRan, Inc., which was facilitated by Ronald Grossman. We both also benefited from the financial support of the Filomen D'Agostino and Max E. Greenberg Research Fund at the New York University School of Law. The funders did not participate in the selection of the topics and authors, or in any editorial decisions.

We are grateful to the unstinting encouragement of Dean John Sexton and the brilliant logistical support of Bobbie Glover, the Law School's Director of Special Occasions, in making the conference a success. We wish to thank Resources for the Future and particularly Paul

Portney for agreeing to publish this book. The various chapters were strengthened by the comments of the anonymous referees, as well as by the important suggestions of Paul Portney, Richard Getrich, and Betsy Kulamer, and the editing of Eric Wurzbacher, assisted by Amie Jackowski. Our able secretaries, Barbara Ortiz and Evelyn Palmquist, contributed to every stage of this endeavor. As always, our families provided us with encouragement and support.

Richard L. Revesz
Richard B. Stewart
New York University

PART I:
Introduction

1

The Superfund Debate

*Richard L. Revesz and
Richard B. Stewart*

During the last decade, the Superfund approach to environmental liability and remediation has become highly controversial. The costs of remedying the environmental problems caused by hazardous substances are great, although Superfund is far from the most costly U.S. environmental program. Its annual costs are in the range of $3 to $5 billion—a fraction of the costs of the federal air or water pollution regulation programs.

Much of the controversy generated by Superfund stems from its far-reaching statutory system of liabilities, which goes far beyond that of the common law.
- Liability is strict; no showing of fault or negligence on the part of a defendant is required.
- Liability is retroactive, in the sense that deposits of waste that occurred before Superfund's enactment can form the basis for liability for remedial costs incurred after its enactment.
- Liability is also joint and several; unless a defendant can show that the risk or harm attributable to it is "divisible," each of the defendants in some way responsible for the wastes at a site can be potentially singled out to bear all of the cleanup costs.

The broad net of Superfund liability includes current and past owners and operators of waste sites and waste generators and transporters. Defendants at Superfund sites include not only large industrial firms, but also a broad array of other entities—municipalities, local dry cleaners, hospitals, and a myriad of small businesses. As a result of this expansive liability regime, Superfund also has had significant effects on the real estate, banking, and insurance industries, as well as on the legal profession.

Defendants have criticized the cleanup levels demanded by the U.S. Environmental Protection Agency (EPA) as excessively stringent and costly. Superfund is also widely regarded as a wasteful and inefficient program, plagued by high transaction costs, serious administrative deficiencies, and long delays in cleaning up sites.

After a contentious debate, the 103rd Congress failed to reauthorize Superfund in 1994. The debate will be renewed in the 104th Congress; in light of the changed political composition of the new Congress, significant amendments may well be adopted. Critics of the current system have urged major changes. The most far-reaching could replace the current system of liability with a tax-funded public works program. Many environmental groups, however, strongly defend the basic features of the Superfund program as essential to fund the cleanup of past hazardous waste problems and to provide strong incentives to prevent them from recurring in the future. Legislative changes to the current program are likely to focus on moderating required cleanup levels, reducing the scope of liability, limiting retroactivity, and creating an expanded cleanup fund financed by insurers.

The purpose of this book, based on the papers presented at the Conference on Superfund Reauthorization: Theoretical and Empirical Issues, which took place at the New York University School of Law on December 3–4, 1993, is to provide a serious look at the issues most relevant to the reauthorization debate and the future of the program. To aid the reader to this end, we introduce in this chapter the relevant components of the Superfund statute itself, including the liability and taxing regimes, the impact of the liability regime on various sectors of the U.S. economy, the site cleanup process, and the determination of cleanup standards.

We then summarize the conference papers (now the chapters in this book) and their links to the ongoing public policy debate. The issues addressed are basic to understanding Superfund and will continue to be relevant long after reauthorization. Although the issues are related, each chapter is self-sufficient. Readers interested in particular issues can accordingly limit their attention to the pertinent chapters.

THE SUPERFUND STATUTE

The dangers of unregulated land disposal of hazardous wastes were powerfully brought to national attention in the summer of 1978, when toxic chemicals surfaced in basements and schoolyards in the community of Love Canal, New York. Between 1942 and 1953, the Hooker Chemical Company filled an abandoned site (a hydroelectric channel) with more than 21,000 tons of chemical wastes. In 1953, Hooker covered the site with earth and clay and sold it to the Niagara Falls Board of

Education for $1. A school and playground were built on the site. The surrounding vacant land was developed into a residential community.

In the years that followed, residents noticed foul odors after heavy rains and during humid conditions, but most attributed these odors to nearby industrial facilities. Increased precipitation in the early 1970s raised groundwater levels, causing thick, oily sludges to seep into basements and accumulate on the surface. In 1976, a joint U.S.–Canada commission responsible for monitoring conditions on Lake Ontario identified high levels of the insecticide Mirex in fish; these contaminants were traced to Love Canal. This finding generated public concern, leading to the initiation of groundwater tests and epidemiological studies. In August 1978, New York's health commissioner declared a public emergency. The news media descended upon Love Canal, broadcasting images of a middle American community mired in a swamp of hazardous waste. The health commissioner's report, "Love Canal: Public Health Time Bomb," coined a powerful metaphor for focusing public attention upon the risks of abandoned hazardous waste sites (Menell and Stewart 1994).

Other contemporaneous events, such as the serious toxic spill of pesticides in the James River in Virginia and the discovery of tens of thousands of barrels of discarded, leaking, and unlabeled wastes in the "Valley of the Drums" in Kentucky, added to the perception of a national crisis. Congress acted quickly. Although it had already enacted in 1976 a comprehensive statute—the Resource Conservation and Recovery Act (RCRA)—to regulate treatment, storage, and disposal of hazardous wastes, RCRA's primary thrust was the prevention of future harms rather than the cleanup of sites inherited from the past.

Hearings began in early 1979, and on December 11, 1980, President Jimmy Carter signed into law the Comprehensive Environmental Response, Compensation, and Liability Act (CERCLA), which came to be known popularly as the "Superfund" statute. There was broad bipartisan consensus on the immediacy of the crisis and the need for legislation—a consensus that will be difficult to replicate for its reauthorization in 1995. Although Superfund was voted in Congress following the November election, in which President Carter had been defeated and the Democrats had lost control of the Senate, the favorable vote was 274 to 94 in the House and 78 to 9 in the Senate. CERCLA was substantially amended in 1986 by the Superfund Amendment and Reauthorization Act (SARA). It was reauthorized once more in 1990 (as part of the Omnibus Budget Reconciliation Act) without any substantive amendments.

Several of the features of the current statutory scheme are particularly relevant as background to the consideration of the chapters in this book. We will focus on the liability and taxing regimes, the cleanup process, and the determination of cleanup standards.

The Liability and Taxing Regimes

Under Superfund, the cleanup of hazardous waste sites is funded by two separate sources: a liability regime and a taxing regime.

The Liability Regime. Superfund contains an extensive and far-reaching liability scheme. Liability is triggered when the government or a private party incurs response costs in dealing with a release or threatened release of hazardous substances into the groundwater, surface water, soil, or air. In the event of such a release, the following categories of parties, often referred to as *potentially responsible parties* (PRPs), are liable: the current owner or operator of the site at which the release occurs; prior owners and operators during whose period of ownership there was disposal of hazardous substances at the site; generators of the hazardous substances; and transporters of the hazardous substances who had responsibility for selecting the site. Liability is also imposed on an owner of the site who, though not otherwise liable, obtains knowledge of the release or threatened release and subsequently transfers the property without disclosing such knowledge.

The liability standard under the statute is strict liability, rather than negligence. Thus, a PRP cannot avoid liability by showing that it met the regulatory or common law standards of care applicable at the time that it engaged in the activity, or even that it was also complying with hazardous waste regulatory standards currently in force.

Liability under Superfund is both retroactive and prospective. In addition to imposing liability for cleanup costs attributable to generation, transportation, treatment, storage, or disposal of hazardous substances undertaken before the passage of the statute, it also attaches cleanup costs for wastes disposed after the passage of the statute.

The defenses to liability are extremely limited. A PRP can escape liability only if it can show that the release or threatened release was caused solely by an act of God, an act of war, an act or omission of a third party, or a combination of these causes. Not surprisingly, only the third-party defense has been of practical significance, and the Superfund statute imposes significant limitations on it. Although this defense is the source of considerable litigation, very few PRPs have successfully established it.

A PRP seeking to defend on this ground must show that the third party was the sole cause of the harm, and that the third party was not the PRP's employee or agent. Moreover, these acts or omissions cannot occur in connection with a direct or indirect contractual relationship between the third party and the PRP asserting the defense. Thus, for example, a generator cannot raise the defense if the need for a cleanup arose as a result of the actions of either its transporter or the operator of the site where the wastes were eventually deposited. The party raising the defense must also show that it exercised due care with respect to the hazardous substances and that it took precautions against foreseeable acts or omissions of the third party.

Moreover, the courts have held that PRPs are jointly-and-severally liable if the harm at the site is indivisible—that is, if the wastes are sufficiently commingled that it is not possible to determine which wastes were responsible for the release. PRPs have the burden of showing that the waste and corresponding cleanup costs for which they are responsible are divisible from those attributable to other parties.

Joint-and-several liability is coupled with a *right of contribution*, so that if one PRP had to pay the full cleanup costs at a site, it could require other PRPs to pay their equitable shares of the liability. The right of contribution, however, is unavailing if other PRPs are insolvent or cannot be located. Accordingly, the solvent PRPs, as a group, must absorb the "orphan" shares of insolvent or absent PRPs. The existence of joint-and-several liability is especially significant in the Superfund context. Because significant periods of time—often several decades—can elapse between the disposal of hazardous substances and the cleanup, it is particularly likely that some PRPs will not be found or will be insolvent once they are found.

The Superfund statute also has causation requirements that are highly attenuated. Thus, a PRP can be liable even if it cannot be shown that its hazardous substances were the ones implicated in the release or threatened release that gave rise to cleanup costs. Liability can be imposed upon a PRP if its hazardous substances at some point were present at a site at which there was later a release or threatened release of the same or even another hazardous substance.

In addition to its cleanup provisions, CERCLA authorizes federal, state, and Indian tribe authorities who manage or control natural resources to sue for damages to such resources resulting from the release of a hazardous substance. The categories of persons liable are the same, and the principles of liability are generally the same, as in the cleanup program. Thus, for example, contamination at a site might lead to the impairment or destruction of wetlands. Following a cleanup that removes the hazardous substances from the site, the PRPs could remain liable for *natural resource damages* (NRD) in connection with any residual damage to the wetlands.

The Taxing Regime. The Superfund taxing provisions are an adjunct to the liability scheme. Currently, three separate taxes are levied on chemicals, petroleum products, and general corporate profits to finance the

Hazardous Substances Superfund (the trust fund), which gives the statute its popular name. This fund is used for two primary purposes: to pay for cleanups at sites at which all the PRPs are either insolvent or unknown, and to advance money for EPA cleanups at other sites pending EPA's recovery of cleanup costs from the PRPs. Thus, the fund is a revolving as well as residual form of financing, which covers cleanup costs that cannot be recovered through the liability scheme. At the time of the passage of CERCLA in 1980, Congress authorized a $1.6 billion fund, with the money to be raised over five years. The 1986 SARA amendments provided for an additional $8.5 billion, also to be raised over five years. When Congress reauthorized the statute in 1990, it provided for a funding level of $5.1 billion between October 1, 1991, and September 30, 1994. The total costs of cleaning up sites potentially subject to Superfund have been estimated at $100 billion or more.

Impact of the Liability Regime

The Superfund liability scheme has transformed vast sectors of the U.S. economy and has had effects far beyond the PRPs at Superfund sites. Our discussion in this regard focuses on the real estate, banking, and insurance industries, as well as on municipalities and the legal profession, where the impact of Superfund has been particularly significant.

The Real Estate Industry. Purchasers of real estate face the threat that they will buy contaminated land and that, at some point in the future, they will face liability as the current owners of the land. CERCLA recognizes an "innocent landowner" defense to owner liability. In order to assert this defense, an owner must establish that at the time it acquired the facility it did not know and had no reason to know about the hazardous substances responsible for the release or threatened release. Having "no reason to know" is further defined as undertaking "all appropriate inquiry... consistent with good commercial or customary practice." In order for purchasers to take advantage of this defense and avoid potentially far-reaching liabilities, it is now customary in the context of transfers of commercial real estate for purchasers to undertake environmental assessments, which, depending on the circumstances, can include extensive testing of soil and groundwater.

The Banking Industry. With respect to the banking industry, the statute, somewhat confusingly, exempts from liability "a person, who, without participating in the management of a [site], holds indicia of ownership primarily to protect his security interest." Two categories of cases involving mortgage lenders are relevant. First, if the borrower defaults and the lender forecloses, taking title to the property, logic would suggest that because the lender then acquires full indicia of ownership it can no longer qualify for the exemption and would face liability as a current owner. Second, preforeclosure liability arises when the bank becomes sufficiently involved in the activities of the debtor—for example, by monitoring the debtor's operations—that it is deemed to participate in the debtor's management. Unfortunately, the courts have been quite divided about what constitutes too much involvement, and a regulation by EPA attempting to clarify the issue was recently struck down in the courts as beyond EPA's authority.

As a result, banks routinely perform, or require the performance of, environmental assessments before they approve mortgages for commercial real estate. Moreover, critics of Superfund claim that the potential liability of banks, and, perhaps more importantly, the considerable uncertainty surrounding the scope of such liability, has undesirably increased the cost and reduced the availability of credit, especially for small businesses.

The Insurance Industry. The insurance industry also has been centrally affected by the Superfund statute. Between 1973 and 1986, the standard *comprehensive general liability* (CGL) policy held by individuals and corporations included a pollution exclusion clause, which provided that insurance would not cover bodily injury or property damage arising out of pollution except if the release was "sudden and accidental." In large part as a result of Superfund, insurers amended this clause in 1986, explicitly excluding any pollution-related liability. The impact of Superfund on the insurance industry is manifested in two distinct ways.

First, firms interested in protecting against liability for pollution must now purchase specialized insurance, which, to the extent it is available at all, carries high premiums, high deductibles, high coinsurance rates, and low caps. Moreover, the availability of such insurance is quite limited. Thus, many firms have had little option but to self-insure, sometimes risking bankruptcy in the event of an environmental accident.

Second, Superfund has raised an enormous amount of contentious litigation concerning the liability of insurers under policies written before the 1986 change in the pollution exclusion clause. PRPs in Superfund actions routinely seek indemnification from their CGL insurers. The interpretation of insurance contracts is a matter of state law, and the state supreme courts that have addressed the issue have split almost evenly on whether the release or threatened release of hazardous substances at Superfund sites is "sudden and accidental" and meets the other terms of policy coverage. The litigation on this matter between insureds and insurers has consumed exceedingly high transaction costs and has led to proposals for the establishment of a fund (to supplement the existing trust fund), financed by assessments on insurance companies, to pay for a portion of Superfund cleanup costs in place of case-by-case litigation between insureds and insurers.

Municipalities. Municipalities also have been caught in the Superfund web. Typically, municipal solid waste contains a small percentage of hazardous substances. Some municipalities disposed of this waste at sites also used by industrial generators. If liability is apportioned proportionally to the aggregate amount of waste contributed by each PRP, on the premise that cleanup costs are roughly proportional to the volume of waste to be cleaned up, the municipalities will generally bear a high percentage of the costs. If, instead, the relevant criterion is the amount of hazardous substances in the waste contributed by each PRP, the bulk of the burden will be placed on the industrial generators. Judicial decisions adopting the former approach have threatened to imperil the financial stability of some small towns.

The Legal Profession. The legal profession has also been powerfully affected by the Superfund liability scheme. In the 1970s, the bulk of environmental law practice consisted in large part of challenging EPA and state command-and-control regulations and the implementation of those regulations. Typical lawsuits pitted industrial firms or environmental groups on one side against the federal government on the other. The specialized environmental bar was disproportionately located in Washington, D.C. Largely as a result of Superfund, environmental disputes now routinely involve controversies among industrial and commercial firms that are PRPs at the same site, and between such firms and insurers and banks. The federal government is sometimes both the enforcer of the law and a polluter responsible for cleanup costs at a site. As a result of the broad scope of Superfund liability, environmental law has become a standard component of legal practice nationwide.

The Cleanup Process

The process leading to the cleanup of Superfund sites is cumbersome and slow, and consists of several stages. First, EPA must become aware of a site's existence. Generally, a site is brought to the agency's attention by a state or municipality, or by citizen complaints; there is no federal discovery program. EPA then places the site in the CERCLA Information System (CERCLIS)—the inventory of locations that potentially require cleanup. To date, over 36,000 sites have entered the CERCLIS database.

Second, EPA conducts a Preliminary Assessment (PA) to ascertain the risks posed by the site. If warranted, a Site Inspection (SI) then follows. At each of these stages, many sites are classified as sufficiently harmless to warrant no further attention.

Third, EPA ranks sites under the Hazard Ranking System (HRS). The HRS is composite score that measures the risk of the site by reference to three possible routes of human exposure: groundwater, surface water, and air.

Fourth, sites that receive a score above a given cut-off are placed on the National Priorities List (NPL); currently there are over 1,200 sites on the NPL. Only sites listed on the NPL are eligible for the expenditure of money by EPA for long-term remedial action from the trust fund. This limitation, however, does not apply to EPA removal actions (quicker and less extensive measures often undertaken in the face of emergencies).

For sites on the NPL, the fifth stage of the process involves the preparation of a Remedial Investigation/Feasibility Study (RI/FS). This stage consists of a more detailed examination of the site and a preliminary study of possible remedies.

Sixth, EPA issues a Record of Decision (ROD). This document contains an analysis of alternative remedies, with their expected costs, and selects the remedy that will be implemented at the site.

Seventh, comes the Remedial Design/Remedial Action (RD/RA). The former is a more detailed design of the remediation technique chosen in the ROD; the latter is the actual cleanup of the site.

The process, however, does not always occur in this linear fashion. Cleanup activities at a site are often divided into separate parcels known as operable units; one unit, for example, might involve soil removal and another, groundwater treatment. The different operable units may progress at different rates: one might be at the RD/RA stage whereas the other might be at the RI/FS stage.

A RAND study completed in 1989 (Acton 1989) showed that, for a site that ultimately gets listed on the NPL, it takes on average forty-three months between the time EPA becomes aware of a site's existence and its listing. Twenty months then elapse until the beginning of the RI/FS, thirty-eight additional months until the issuance of the ROD; the RD/RA takes an additional forty-three months. Thus, on average, the time elapsed between listing on CERCLIS and the completion of the RD/RA is twelve years; eight-and-a-half years elapse between the listing on the NPL and the completion of the RD/RA.

Typically, EPA or a state in which a site is located is responsible for the stages leading to listing on the NPL. Of the later stages, the RI/FS and the RD/RA can be conducted by EPA or the state, or by a group of PRPs. In contrast, the issuance of the ROD is the sole responsibility of EPA.

In the early years of the Superfund program, EPA followed a "fund lead" strategy for cleanup. It hired contractors to carry out cleanup activities, paid them out of the fund, and then sought reimbursement from PRPs. The limited size of the fund and the delays and difficulties in obtaining reimbursement led EPA to make increasing use of an "enforcement lead" approach. Under this approach, EPA uses its CERCLA authority to issue an administrative order to PRPs or seek a court order requiring the PRPs to undertake the cleanup. Currently, PRPs are undertaking the bulk of RI/FSs and RD/RAs, typically as a result of settlements with EPA. Evidence suggests that the cost of a given cleanup is about 20% lower when it is undertaken by the PRPs rather than by EPA (see Chapter 6 in this book), presumably because private PRPs have stronger incentives to minimize costs and can supervise contractors more effectively.

The Determination of Cleanup Standards

The most important decision at any NPL site is the determination of the extent of the cleanup and the choice of cleanup technology. If the site's soil is contaminated, should the site simply be capped to reduce the probability of releases into the groundwater, or should the soil be removed and incinerated off-site? The first option will typically be a great deal cheaper, but might pose some long-term risks. Similarly, in the face of groundwater contamination, is it sufficient to prevent migration of the contaminated groundwater through containment measures and secure an alternative source of drinking water (or do nothing at all if the contaminated groundwater is not used for drinking), or instead should one undertake a "pump and treat" program? The latter course of action will be far more expensive, and there is substantial question about its long-term effectiveness. Unfortunately, the statute says little that is helpful in answering these questions, and a wide range of remedies has been used in actual cleanups at NPL sites.

CERCLA contains two sets of provisions dealing with cleanup standards. First, it directs EPA to select remedies protective of "human health and the environment." In making this determination, EPA is to consider a wide range of factors. For example, remedial actions must be "cost effective," but must also "to the maximum extent practicable" utilize "permanent solutions" and technologies that will result "in a permanent and significant decrease" in the volume, toxicity, and mobility of contaminants. These provisions leave EPA with considerable discretion. EPA has tended to emphasize more permanent and costly remedies, such as treating contaminated groundwater rather than simply taking steps to prevent its migration.

Second, CERCLA requires that sites be cleaned in accordance with any "legally applicable" or "relevant and appropriate" standards (ARARs or applicable or relevant and appropriate requirements), where such standards exist. Any standard promulgated under a federal environmental law is "legally applicable" and therefore automatically an ARAR; more stringent state standards are ARARs, if certain procedural conditions are met. The statute, however, does not define when standards are "relevant and appropriate," and therefore also ARARs.

The ARAR prescription is particularly problematic in the case of groundwater contamination. The Safe Drinking Water Act (SDWA) defines permissible levels of various pollutants in publicly supplied drinking water. If groundwater at a Superfund site is contaminated by such a pollutant, the SDWA standard probably will not be deemed "legally applicable" if this groundwater is not used as the source of publicly supplied drinking water, or if it is treated before its distribution to households. The standard might, however, be deemed "relevant and appropriate" and therefore qualify as an ARAR. The Superfund statute provides that where SDWA standards are "relevant and appropriate," the cleanups must at least achieve the Maximum Contaminant Level Goals (MCLGs) under the SDWA. (The SDWA provides that MCLGs "be set at the level at which no known or anticipated adverse effects on the health of persons occur and which allows an adequate margin of safety." MCLGs are aspirational goals; the enforceable standards or Maximum Contaminant Levels (MCLs) must be set as close to MCLGs "as is feasible.") In the case of known or probable carcinogens, MCLGs require a zero concentration of pollution—probably an unattainable objective in groundwater remediation.

The requirement that Superfund cleanups satisfy ARARs, however, is subject to significant exceptions. In the case of groundwater remediation, ARARs need not be used in a variety of circumstances, including if "the remedial action includes enforceable measures that will preclude human exposure to the contaminated groundwater." More generally, for any remediation financed solely by the trust fund, exceptions from ARARs are appropriate on the basis of a balance between "the need for protection of public health and welfare and the environment at the facility," and the availability of amounts from the trust fund to respond to other sites. Moreover, ARARs generally do not exist for soil remediation, which is a major element in many cleanups.

SUPERFUND TODAY: THE STATUS OF THE DEBATE

Superfund is today the environmental program that practically everyone loves to hate. Industrial firms, banks, and municipalities complain about the breadth of the liability regime and the high cost of cleanups. Insurers complain about being subjected to claims that they believed they had long ago contracted out of. Small businesses complain about high transaction costs. Traditional environmental groups complain about the slow pace of cleanups and the small number of sites where the remedial action has been completed. Environmental justice groups complain that disproportionately few cleanup resources have been deployed in poor and minority communities.

Despite the strong feelings that it has engendered, the Superfund program has been the subject of little dispassionate study. This book brings together some of the most recent important theoretical and empirical work on four issues central to the evaluation of Superfund: cleanup standards, the liability regime, transaction costs, and natural resource damages. The first three are the most salient issues in the reauthorization debate. The fourth has, until now, received less attention, yet it is potentially no less significant. Taken together, the chapters in this book, which are written from a variety of intellectual perspectives, present a mixed assessment of the Superfund program: they portray it as neither a model of successful public administration nor the monster that it is often alleged to be.

Cleanup Standards

In addition to the general recognition that the current statute provides imprecise and contradictory commands for the determination of cleanup standards, most critics of Superfund believe that cleanup costs are too high in light of the corresponding benefits. The figures are, indeed, staggering. The average cleanup cost at an NPL site is currently over $30 million. A recent study by the Congressional Budget Office places the total cleanup costs for current and future NPL sites in the $100 to $400 billion range.

The three chapters on this topic consist of empirical studies of RODs that seek to ascertain which factors explain EPA's cleanup decisions. The chapter by Katherine Walker, March Sadowitz, and John Graham presents ten widely held stereotypes about Superfund cleanups and shows that these myths generally do not hold true. Several of their findings are particularly noteworthy. First, both the carcinogenic and noncarcinogenic risks at most NPL sites, as assessed by EPA, are substantial, and even if these risks were being systematically overstated by considerable amounts, remedial action would nonetheless be appropriate; this finding contradicts the view that Superfund expenditures are typically used to reduce already trivial risks. Second, however, estimated health risks at sites are not adjusted to reflect actions that might be taken to restrict access to the site or to contaminated groundwater; while this finding is consistent with the statutory preference for more permanent remedies, it cannot be justified in cost-benefit terms since fencing a site or providing alternative water supplies is generally far cheaper than treating soil and groundwater. Third, remedial decisions do not take into account the number of people affected at a site; this finding is consistent with the statutory requirement that ARARs be used but is again inconsistent with the prescriptions of cost-benefit analysis, which would call for more stringent cleanups at sites that affected larger populations.

In their chapter, James Hamilton and W. Kip Viscusi also conclude that many of the estimated hazards at Superfund sites are substantial. They find, however, that the bulk of the risk that forms the basis for these estimations results from exposures to future populations that might be located on or near the site as a result of changes in current land use patterns. Most strikingly, the authors note that about 80% of the carcinogenic risk at Superfund sites is to residential populations that are hypothesized to move to Superfund sites in the future.

The Hamilton and Viscusi findings should lead one to question whether one should engage in an expensive, permanent cleanup rather than taking the far less costly steps of containing the contamination, erecting a fence, and placing use restrictions in the property's deed. While attractive at first glance, the nonpermanent alternative nonetheless might be deemed unsatisfactory for a variety of reasons. For example, one might be skeptical that the deed restrictions will be respected in the future, when public attention might have shifted to a different pressing social problem. Moreover, if a permanent cleanup is not performed, extensive operation and maintenance activities will typically have to be carried out indefinitely to ensure that the contamination does not spread. Under the CERCLA statute, the states, rather than EPA, have operational and financial responsibility for these activities and might lack the interest or expertise to carry out their responsibilities effectively. Also, the surrounding communities might object vigorously to the presence of vacant contaminated land in their midst. Because of the current perception that communities surrounding NPL sites are often disproportionately poor and minority—an issue that is the subject of ongoing empirical research—the standards for cleanup are a matter of particular concern to the environmental justice movement. Nonetheless, the findings of the Hamilton and Viscusi chapter raise important questions that merit serious consideration.

The final chapter in this section, by Shreekant Gupta, George Van Houtven, and Maureen Cropper, seeks to determine whether the benefits of long-term cleanups are worth their costs. Because the information currently available is not sufficient to conduct a traditional cost-benefit analysis, the authors ask whether a comparison of costs and benefits played a role in the choice of target risk—the lifetime risk of death as a result of contamination remaining at the site after cleanup—as well as in the choice of treatment technology. The cost-benefit criterion would dictate acceptance of higher target risks at sites with higher cleanup costs. The authors find that EPA did not accept higher target risks at such sites, following the goal of protecting health without considering costs. The authors also test the claim of the environmental justice movement that EPA chooses higher target risk levels in areas that are disproportionately poor or minority. When adjusting for other relevant factors, the authors reject this hypothesis.

Underlying these three chapters is the question of whether and to what extent the cost-benefit criterion should play a role in cleanup decisions—an issue that has received close attention in the reauthorization debate. Opponents of Superfund argue that it is wasteful to engage in equally extensive cleanups regardless, for example, of the population affected by the site. Similarly, they maintain that it is irrational to insist on permanence without regard to the additional costs of more permanent cleanup strategies. If the objective were to maximize social welfare, their criticisms would certainly be well taken. However, a well-established tradition in many environmental regulatory programs holds that standards should be set at the level necessary to protect public health without considering costs, so that every person should be guaranteed some minimum level of environmental protection. The exercise is, to be sure, somewhat unreal and commentators have suggested that EPA, barred from considering costs explicitly, does so implicitly. It is nonetheless useful to understand that Superfund is not an aberration in this regard.

The Liability Regime

The liability regime also figures prominently in the reauthorization discussions. PRPs and insurers attack its breadth. In particular, they claim that joint-and-several liability is unfair because it makes solvent PRPs pay for their insolvent counterparts, and that retroactive liability fails to create any desirable incentives. They would prefer nonjoint liability, or, even better, the scrapping of the liability system and its replacement with a tax, ideally a generally applicable tax, to fund cleanups.

The chapter by Lewis Kornhauser and Richard Revesz provides a theoretical analysis of the relative merits of joint-and-several liability as compared with nonjoint liability. They employ three criteria: deterrence, settlement-inducing properties, and fairness. They conclude that, with respect to each of these criteria, neither rule dominates the other. While their analysis proceeds from a simplified economic model and does not explicitly account for the many real-world complications that arise in every Superfund case, its major insights are applicable to more complex situations. Kornhauser and Revesz conclude that the burden is on opponents of joint-and-several liability to justify a departure from the status quo and that they have failed to show that nonjoint liability would be superior.

The chapter by Katherine Probst addresses a similar question from a different perspective. As described above, the trust fund acts as a residual source of funding for cleanups, providing financing at sites where the liability regime fails to raise sufficient money. Thus, for a given set of cleanup standards, any cutbacks on the liability must be counteracted by a tax increase to finance the trust fund. Probst examines the financial impact on various sectors of the economy of three different Superfund policies: the status quo; a waiver of liability for wastes disposed of before 1980 at multiparty sites; and the Clinton administration's proposal, which, in certain circumstances, would apportion liability shares among PRPs and have the trust fund, rather than the solvent PRPs, pay for the shares of insolvent, though not of unidentifiable, PRPs.

Probst finds that the relative percentage of total cleanup costs borne by each industry remains almost constant under each liability option. She estimates that the chemical and allied products industry bears the largest percentage of cleanup costs, about 25%. Even for this industry, however, the burden of Superfund liability is relatively small under any of the alternatives. Probst acknowledges, however, that Superfund expenditures may place a significant burden on some less profitable industries, such as mining and wood preserving, and on individual firms.

Transaction Costs

The Superfund program has been sharply and persistently criticized as involving excessive *transaction costs*—costs incurred in the process of determining cleanup remedies and imposing financial liability—that do not contribute to the cleanup process. Echoing a widely held sentiment, President Bill Clinton has suggested that more is being spent to enrich lawyers than to clean up the environment, and he invoked this failing as a principal justification for amending CERCLA. It is widely believed that the Superfund program has become mired in endless and costly wrangles among legions of lawyers representing EPA, scores of PRPs, and insurers.

Building on its empirical research on the civil justice system, RAND has gathered data on the transaction costs of Superfund. Lloyd Dixon's chapter summarizes the results of this research. He finds that private sector transaction costs are indeed significant, ranging from roughly 23% to 31% of all private sector Superfund liability outlays. The data also confirm that transaction costs are much higher at sites with large numbers of PRPs than at sites with one or a few PRPs. Finally, the data show that transaction costs are a much higher percentage of total outlays for PRPs that contributed small shares of waste at a site.

Are Superfund transaction costs excessive? As Dixon points out, the answer depends in part on comparative judgments. As a percentage of total outlays, private sector transaction costs under Superfund are somewhat less than defendant transaction costs as a percentage of outlays in tort litigation generally (35%) and much less than asbestos claim litigation (50%), although higher than airline crash litigation (14%). But are these the right benchmarks for evaluating Superfund?

For one thing, RAND's research did not examine the transaction costs incurred by the government as plaintiff. In ordinary civil litigation, the plaintiffs' transaction costs are roughly comparable to those of defendants. In the case of Superfund, however, there is one dominant "repeat player" plaintiff, EPA, which should be able to enjoy substantial scale economies and hence lower transaction costs. (To a lesser extent, the same ought to be true for the states or large PRPs, which are the plaintiffs sometimes, though far less frequently.) Accordingly, if plaintiff transaction costs were included in the analysis along with defendant costs, one would expect that total Superfund transaction costs as a percentage of total outlays would be lower, compared to the percentages for tort litigation, than RAND's figures indicate.

On the other hand, there are good reasons for supposing that Superfund transaction costs should be much lower than the general costs of civil tort litigation. As noted, EPA is the plaintiff in almost every case. Moreover, many disputed issues are resolved through an EPA administrative process, subject to limited judicial review on the administrative record, rather than by de novo civil trial. Finally, the sweeping rules of liability under CERCLA eliminate or minimize the need to resolve many of the factual issues presented in tort litigation, such as fault and causation. All of these factors should result in much lower transaction costs for Superfund. They suggest that an alternative benchmark might be workers' compensation systems, where the transaction costs of the lowest-cost administrative systems can approach 20% or less, or the social security disability system, where transaction costs are less than 10%. These areas, however, lack the technical or factual complexities that typically arise in Superfund cases. Also, they have well established case law, whereas under Superfund, because of the relative recency of the program, some basic legal issues are still unresolved and must be litigated.

Dixon's finding that PRPs with a small share of the liability bear a disproportionate amount of transaction costs provides the point of departure for the other chapter in this section: Lewis Kornhauser and Richard Revesz's study of *de minimis settlements*—settlements with parties responsible only for a small share of the liability at a site.

Congress, aware of this disparity in transaction costs, prescribed in the 1986 SARA amendments that "[w]henever practicable and in the public interest," EPA "as promptly as possible" enter into settlements with de minimis PRPs. Kornhauser and Revesz conducted an empirical study of EPA's use of de minimis settlements. They found that EPA has vastly underutilized this tool. Even when it has entered into such settlements, it has done so late in the cleanup process, after years of legal wrangling have greatly reduced the benefits of settlement. It has also failed to follow its own policy of standardizing the form of the settlements, thus creating incentives for costly negotiations over the terms of de minimis settlement and for conflict between de minimis and non–de minimis defendants.

Throughout the country, thousands of small businesses are telling essentially the same dispiriting tale. Their share of the liability at a site is expected to be only a few thousand dollars; they are prepared to pay a hefty premium to compensate EPA in the event that the initial estimate of remedial costs is too low; and they would like to send EPA a check to settle the case as soon as possible. Instead, they are told to wait, either until all de minimis PRPs have organized into a committee or until the major PRPs are ready to undertake a cleanup. Years go by; the de minimis PRPs feel compelled to have at least some sort of legal representation; legal costs soon overtake their expected liabilities; and their access to credit markets is imperiled.

How can this situation be improved? The answer is not easily determined because EPA already has appropriate settlement tools, and Congress has made reasonably clear its interest in the entry of such settlements; repeating the same thing once again is unlikely to make much difference. Part of the problem is that EPA, as litigant, appropriately concerns itself with maximizing its own expected recovery in litigation and minimizing its administrative costs. As the agency charged with improving social welfare by responding to threats to health and the environment, however, it ought to concern itself with the costs that it imposes on third parties. The problem also stems from EPA's lack of bureaucratic resources as well as of sufficient agency incentives to promote de minimis settlements. Efforts to alleviate the plight of de minimis PRPs will be unsuccessful unless these institutional realities are taken into consideration.

Natural Resource Damages

Attention and debate on CERCLA has focused on liability for cleanups. The natural resource damages (NRD) program is less fully developed, has not resulted in comparable liabilities or litigation, and thus far has not been a significant issue in the reauthorization debate. Nonetheless, NRD liability is a rapidly expanding and potentially quite significant aspect of CERCLA. It must be reckoned with in any comprehensive reevaluation of the Superfund program.

The NRD regime, according to Richard Stewart's chapter, is a novel blend of tort liability, public trust, and administrative models. Stewart argues that this hybrid system creates significant conceptual, legal, and practical difficulties, including high transaction costs, wasteful expenditures of recoveries, and severe difficulties in developing appropriate measures of damages. One measure of damages is the cost of restoring the injured resource. There is controversy as to whether restoration requires physical and biological replication of the injured resource, which could be enormously costly, or may include trustee acquisition of other resources providing comparable services to the public.

Another measure of damages is the diminished value of the injured resource. The difficulty here is that market measures of value are likely to be quite inadequate as applied to environmentally significant natural resources. Trustees and some economists have sought to base damages on the results of contingent valuation methodology surveys that ask members of the public how much they would hypothetically be willing to pay to preserve or restore a resource. The validity and reliability of this technique is hotly disputed.

The significance of NRD liabilities is likely to grow, as state and tribal as well as federal trustees become more familiar with the program and the possibility of substantial recoveries, which trustees can use to support their operations. Several pending cases seek damages in excess of a billion dollars. It remains to be seen whether the problems identified by Stewart can be resolved with further experience or whether the NRD program will eventually generate demands for changes analogous to those which now attend the cleanup liability regime. Stewart is pessimistic about the performance of the current NRD program and recommends a simplified system of scheduled damages in place of the current tort-based approach, which requires case-by-case proof of injury, causation, and damages.

There is, moreover, a close link between the choice of cleanup standards and the magnitude of NRD liability. To the extent that EPA pursues ambitious cleanup remedies, directed at removing most or all groundwater or soil contamination in the name of environmental protection, it is effectively engaged in restoration of natural resources. If, however, the current reauthorization leads to a significant relaxation of cleanup standards, state and tribal governments could demand costly restoration measures for natural resources under their management or control, thereby undermining the legislative efforts to reduce the cost of the Superfund scheme. It follows, therefore, that efforts to reform the cleanup program should address its interrelationship with the NRD program.

* * *

The issues discussed in this book are complex and do not lend themselves to one-sided treatment. We are convinced that the political process, which will undoubtedly hear many shrill voices, would benefit from paying close attention to the nuanced theoretical and empirical research presented in the chapters of this book.

More importantly, the issues raised here are certain to remain important even after the Superfund statute is eventually reauthorized, regardless of what form the reauthorization takes. This book will thus remain a valuable public policy analysis of what has become one of our most visible and controversial environmental programs.

REFERENCES

Acton, Jan A. 1989. *Understanding Superfund: A Progress Report.* RAND/R-3838-ICJ. Santa Monica: RAND.

Menell, Peter S., and Richard B. Stewart. 1994. *Environmental Law and Policy.* Boston, Mass.: Little, Brown & Co.

PART II:
The Cleanup Standards

2

Confronting Superfund Mythology: The Case of Risk Assessment and Management

Katherine D. Walker, March Sadowitz, and John D. Graham

Concerns about the current and possible future costs of the Superfund program have sparked debate about the program's decisionmaking process and its ultimate goals. Particular attention has been given to the U.S. Environmental Protection Agency's (EPA) conduct and use of risk assessment in remedy selection. Some argue that EPA's risk assessment techniques exaggerate risks (Hazardous Waste Cleanup Project 1993); others maintain that the same techniques may understate risks in some instances (Finkel 1989). In either case, the assumption is that risk assessment plays a critical role both in decisionmaking and in the amount and ultimate cost of cleanup efforts at hazardous waste sites.

We believe that a Superfund mythology, sprung from the recounting of anecdotal experiences at individual sites, has confused the debate about Superfund reform. In the absence of hard data about a representative sample of Superfund sites, arguments about the broader implications of the anecdotal incidents divert attention from more fundamental scientific and value questions that must be answered to improve the operation of the program. While extensive documentation is publicly available in the administrative records developed for each site, few attempts have been made to assemble, analyze, and publicize findings about a large sample of sites.[1] In particular, the role of risk assessment in the Superfund program needs clarification, a process that would benefit from the availability of comprehensive data.[2]

Hence, the Harvard Center for Risk Analysis (HCRA) has undertaken a rigorous nationwide survey of Records of Decision (RODs) in order to document the official basis for cleanup decisions at individual sites and, where necessary, to dispel the myths that have developed about the Superfund program. This chapter focuses on myths surrounding the human health risks reported at Superfund sites and explores the implications of HCRA findings for Superfund reform.

SUPERFUND RISK ASSESSMENT AND MANAGEMENT

The original Superfund law—the Comprehensive Environmental Response, Compensation, and Liability Act of 1980 (CERCLA)—required EPA to take action at hazardous waste sites that pose "a substantial endangerment to public health or welfare or the environment" [CERCLA Section 104(a)]. EPA chose to use risk assessment techniques, then in their infancy, to inform decisions about when and how much to clean up.

The "baseline risk assessment," an evaluation of hazards that the site poses or may pose in the absence of remedial action, is used to buttress the decision whether or not to take remedial action. It may also be used in the selection of "cleanup goals" at the site: the levels to which specific contaminants must be reduced to protect public health, welfare, and the environment. Finally, a form of comparative risk assessment is used sometimes to compare remedial alternatives on both the basis of their abilities to meet the cleanup goals and their tendencies to generate risks in implementation. In this study, we focus on the first two uses of risk assessment at Superfund sites because comparative risk assessments of remedial alternatives are generally qualitative and often cursory.

Superfund Risk Management Criteria

EPA policy on what levels of risk warrant remedial action has evolved over the last decade. In the early years of the program, a cancer risk of 1×10^{-6} (a chance of one in one million of contracting cancer) was often informally used as the benchmark against which all estimated risks were judged; risks lower than 1×10^{-6} were considered negligible but risks higher were presumed eligible for remediation.[3] The National Contingency Plan (NCP), the regulations implementing the original Superfund law and its subsequent 1986 amendments (Superfund Amendments and Reauthorization Act, or SARA), designated an acceptable "risk range" of 1×10^{-6} to 1×10^{-4}. EPA site managers have some flexibility in deciding whether to remediate a site when estimated cancer risks fall within this range. In mid-1991, EPA's Office of Solid Waste and

Emergency Response published a policy directive indicating that risks within the range generally should not be remediated unless adequate justification is given (U.S. EPA 1991b). Consequently, many of the sites with remedies selected prior to 1992 may reflect earlier policies when risks less than 1×10^{-4} were more likely to be remediated.

Even though a formal written policy for noncancer hazards does not exist, EPA may decide to remediate when the hazard quotient or the hazard index is greater than 1.[4] When the hazard quotient exceeds 1, it indicates that the intake of a chemical exceeds its *reference dose* (RfD). EPA defines the RfD as "an estimate (with uncertainty spanning perhaps an order of magnitude or greater) of a daily exposure level for the human population, including sensitive subpopulations, that is likely to be without an appreciable risk of deleterious effects during a lifetime" (U.S. EPA 1989). Given the threshold mechanism by which most noncarcinogens are believed to act, a hazard quotient greater than 1 is presumed to indicate an increased likelihood of an adverse effect. However, the magnitude of the value is not a direct measure of either the likelihood of adverse effects (as it is in the models of cancer risk assessment) or of the potential severity of the impact. A hazard quotient less than 1 is presumed to be safe or without likelihood of an adverse effect. The same interpretation applies to the hazard index.

In the discussion that follows, it is important to keep in mind that numerical risk management criteria are not absolute and that, therefore, the significance of the levels of risk reported at hazardous waste sites may be viewed differently over time. Although widely used in federal and state environmental programs, numerical risk management criteria are rarely mandated by statute; change over time for inexplicable, often political or bureaucratic reasons; and rarely have been based on careful policy analysis. In a previous study of the risk management criteria used by EPA for regulating carcinogens, HCRA found that inadequate consideration was given to key policy considerations, such as economic efficiency and cost-effectiveness, various notions of equity, and the quality and weight of scientific evidence. This same study also provides a regulatory history of how numerical risk management criteria were exported from the U.S. Food and Drug Administration to EPA (Rosenthal, Gray, and Graham 1992).

It is equally important to recognize that numerical risk management criteria for the protection of public health, though often the most visible criteria for decisionmaking, are not the only ones. The Superfund statute also places a strong emphasis on the protection of existing or potential groundwater supplies. This often understated yet sometimes overriding goal frequently exists in apparent tension with efforts to assess potential health risks and select appropriate levels of remedial action on the basis of likely human exposures at a site. This tension reflects a need for the U.S. public and the Congress to have a clear dialogue about the fundamental goals of the Superfund program and the role of risk assessment in achieving them.

THE DESIGN OF HCRA'S SUPERFUND DATABASE

HCRA has developed a database of Superfund sites by extracting information from EPA's official RODs. The database is programmed in DBASE IV and consists of four smaller databases: site identification information, risk information, remedial alternatives, and remedial goals. Each small database is linked by the EPA ROD number, the unique number assigned to each ROD document.

Data from each ROD were collected by research assistants and recorded on standardized data collection forms according to HCRA guidelines. All research assistants received intensive initial training on the proper coding for each question on the form. Readers participated in weekly meetings to discuss and solve problems. Each ROD data collection form was checked by a second reader for completeness and accuracy prior to data entry. All data entries were checked for accuracy prior to analysis.

Calendar year 1991 was selected as the first year to be included in the database because it was the most recent year with RODs available for the full year. The analysis in this chapter is based on 152 RODs from a total of 193 available for selection. Of the 193, we excluded 41 (14 ROD amendments, 23 RODs for federal facilities, and 4 duplicate RODs). Amendments were excluded because they generally present the rationale for changes to the original remedial decision and so often do not have much of the information we were seeking.[5] Federal facilities have not been included at this stage because of technical differences in data collection requirements, but may be the subject of future study. All ten EPA regions are represented in similar proportions to the number of sites from each region on the National Priorities List.

SUPERFUND MYTHOLOGY

While there are many myths about the Superfund program that deserve scrutiny, those surrounding the dangers posed by individual sites to "public health or welfare or the environment" were of particular interest to us. In the discussion that follows, we present ten myths about risk assessment at Superfund sites that served as the basis for our preliminary analysis of 1991 RODs. (As used here, a myth is an item of conventional wisdom that is often grounded more in anecdote or exaggeration than in fact.) Paired with each myth is our finding: a summary of what we found, upon examining each myth, that could and could not be supported by the evidence. We then discuss the particulars of each finding.

The bases for the myths we have articulated would not be able to withstand scientific scrutiny; the myths represented the views of various members of HCRA's scientific staff at the outset of the project, although they may represent broader perceptions as well. Like many myths, however, the myths we discuss contain some element of truth—for instance, they may be true for particular sites—but we found that they are not well founded in the Superfund program as a whole.

Myth 1: **The impacts of Superfund sites on public health or welfare or the environment are all-important considerations in remedial decisions at Superfund sites.**

Finding: **Public health is the primary rationale given for decisions to take remedial action at Superfund sites.**

Given the emphasis placed by the Superfund statute and regulations on the protection of "public health, or welfare, or the environment," it would be reasonable to expect that the three goals would each be given substantial consideration when remedial decisions are made. In reality, protection of public health is the predominant justification cited for the need to take action at the sites we have reviewed to date. Of the 148 RODs for sites at which actions were taken, 94% (139) emphasized concerns about public health impacts only.[6] Three RODs cited both public health and the environment as the basis for action, but only two cited environmental concerns alone as the principal reason for remedial action. None cited welfare or other concerns.

This almost exclusive emphasis on public health reflects EPA's primary focus for the first decade of the Superfund program. Only in the last few years has EPA focused greater attention on potential environmental impacts. Consequently, the assessment of the environmental impacts has been cursory and discussion about such impacts in the RODs limited. The definition of "welfare" impacts remains poorly explained and discussed.

Since public health risks at Superfund sites are stated as the primary rationale for often costly remediation efforts, it is understandable that EPA's health risk assessments have come under increasing scrutiny (Hazardous Waste Cleanup Project 1993). If EPA's health risk estimates are ultimately shown to be grossly exaggerated (an issue that we do not tackle in this chapter), the documented rationale for the Superfund program may prove to be quite slim. It is troubling that the other two statutory triggers for remedial action in the Superfund law—environmental and welfare effects of hazardous wastes—have not been as fully addressed by EPA's decision process (at least as documented in the 1991 RODs). Given the substantial cost of the Superfund program, it is certainly not adequate for EPA to assume that the reported human health risks stand as a reliable surrogate for serious yet unspecified environmental and welfare effects. Since the health risks at sites ultimately may be preventable through relatively inexpensive measures that may not protect the environment or welfare (Hegner 1994), EPA needs to develop analytic procedures for gauging the gravity of nonhealth concerns on a site-by-site basis. Only by understanding the full range of effects of Superfund sites can Congress, EPA, and the public come to an informed conclusion about whether we are spending too many or too few resources cleaning up hazardous wastes.

Myth 2: **If advances in cancer risk assessment showed the risks of cancer at Superfund sites to be negligible, there would be little public health rationale for cleanup.**

Finding: **Both the estimated noncancer as well as cancer risks typically meet EPA criteria for taking remedial action.**

Conventional wisdom has it that cancer risks are the primary or exclusive public health motivation for taking remedial action at Superfund sites. Two factors may account for this impression. One factor is methodological: using current cancer risk assessment methodologies, risk analysts can predict increases in cancer risk at doses generally below those doses at which the potential for noncancer effects begins theoretically to increase. The other factor is emotional: as a society, we have a particular dread of cancer. Cancer risks may yet be primary, but (in light of EPA's current risk management criteria) they do not appear to be the exclusive basis for decisionmaking.

HCRA gathered the maximum cancer risk and hazard index values reported for each site.[7] These maximum values are values specific to an individual exposure scenario, medium, and intake route (for example, residential ingestion of groundwater) as reported in the RODs. To allow comparability between sites, the analysis did not attempt to combine risks across scenarios and media (for example, from residential ingestion of groundwater and dermal contact with contaminated soil), although such combinations were occasionally reported in the RODs.

Figure 1 indicates that about 81% (100) of the reported maximum cancer risks exceeded the "acceptable risk range" of 1×10^{-6} to 1×10^{-4} established in the National Contingency Plan. That is, most of the sites

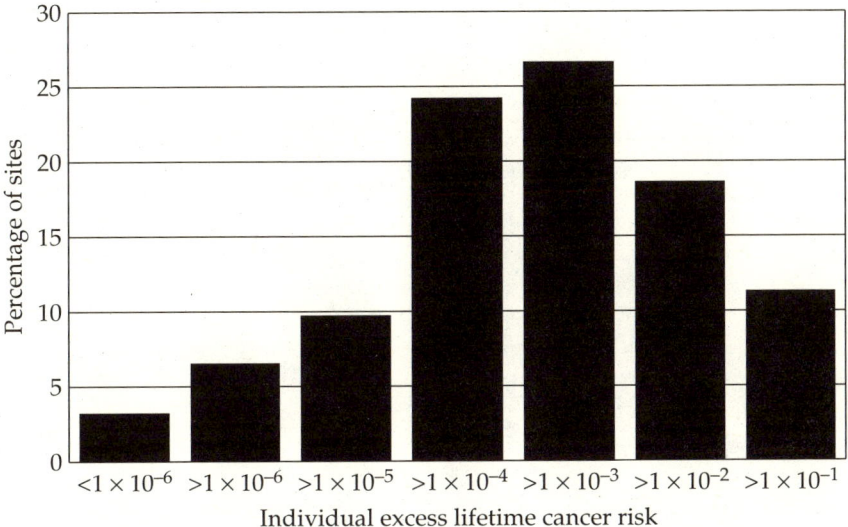

Figure 1. Maximum Individual Cancer Risks at 124 Sites Distributed by Cancer Risk Levels

Note: Maximum values for cancer risks are those reported for any site that reported risks (124/160). These maximum values are specific to individual exposure scenarios, media, and intake routes as reported in the RODs. That is, they do not include cancer risks based on combined exposures for multiple media or intake routes.

reported a cancer risk exceeding 1×10^{-4} for at least one exposure pathway. Approximately 30% of sites with quantified cancer risks reported a maximum cancer risk of greater than 1×10^{-2}.

More surprising to us was the finding, shown in Figure 2, that the maximum hazard indices (for noncancer health effects) were greater than 1 for 74% of the sites and greater than 10 for nearly 50% of the sites. Approximately 18% of the sites reported hazard indices greater than 100.[8]

The number of sites with maximum cancer risks and hazard indices at the high end of the distribution would be greater if risks from different media and intake routes were added for the same exposed individual (for instance, if one added the risk associated with ingestion, inhalation, and dermal uptake of contaminants from soil and potable water for the child who plays in soil and takes a bath before bed).

In a separate analysis, we did find that cancer risks were more often cited as a factor in remedial decisions than were noncancer hazards, indicating that they may indeed be used more often to justify cleanup. However, the magnitude of the hazard indices reported for the hazardous waste sites in the database suggests the need for better understanding of the potential for noncancer health effects. Even if agreement

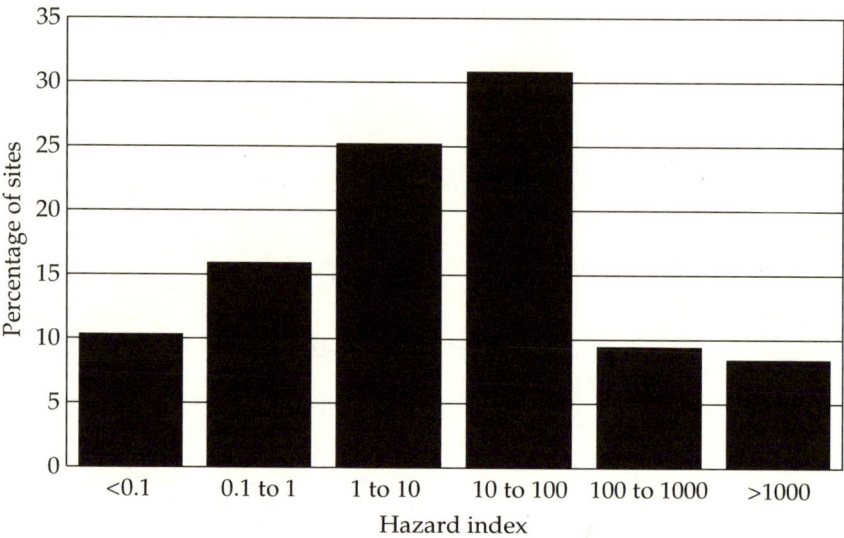

Figure 2. Maximum Noncancer Hazard Index Reported at 107 Sites Distributed by Level of Hazard Index

Note: The level of hazard index is based on the maximum value reported for any site that reported noncancer risks (107/160). These maximum values are specific to individual esposure scenarios, media, and intake routes reported in the RODs. That is, they do not include hazard indices based on combined exposures for multiple media or intake routes.

were reached today to revise guidelines for carcinogen potency assessment in ways that would render negligible any estimated cancer risks at sites, decisionmakers would still be faced with the results of noncancer assessments. Given EPA's risk management criteria, adequate justification appears to exist for action at most sites on the basis of potential noncancer effects alone.

Myth 3: **Correcting errors of understatement or overstatement of risk would radically change decisions at Superfund sites.**

Finding: **Neither increasing nor decreasing the degree of conservatism alone is likely to profoundly influence remedial decisions.**

If one wanted to argue, for example, that no remedial action is necessary at most of the sites in our survey, one would need to establish either that the reported cancer risks and hazard index values are consistently and grossly exaggerated and/or that the risk management criteria used by EPA are too stringent. We have already noted that EPA's risk management criteria are poorly conceived, but it is not clear at a glance whether better-reasoned management criteria would be more or less stringent than those EPA currently employs. On the technical procedures, EPA's own risk assessment guidance for Superfund indicates that both exposure and toxicity values are designed to be conservative so that the estimated risks at sites are not understated. We do not pretend in this chapter to resolve the arguments about the likely degree of conservatism incorporated in Superfund risk estimates, but our findings shed some light on the possible management implications of such understatement or overstatement of risk.

Under EPA's current risk management criteria, adding more conservatism to the risk assessment process is not likely to significantly influence the decision to remediate since most sites already appear to justify remediation with current assessments. If management criteria were to become less stringent, possible sources of insufficient conservatism (such as contaminants without cancer or noncancer toxicity values or the presence of susceptible subpopulations) would need to be considered carefully.

The more salient issue today seems to be the potential for overestimating health concerns and thus triggering remediation when it is not warranted. If risks are currently overstated by a factor of less than five, then correcting the technical estimates would only cause a small fraction of sites to become eligible candidates for "No Action" (again assuming current management criteria). In order for a large fraction of sites to become ineligible for remedial action through more accurate risk assessments, one must show that the toxicologic and exposure assessments, alone or in combination, overstate risks by a factor of 100 or more. This would be enough to move those sites with a maximum risk of 1×10^{-2} to a level below EPA's 1×10^{-4} criterion for taking action. (Similar reasoning applies to the noncancer hazard indices). These generalizations on overstating risks need to be qualified with the observation that a change as small as a factor of two or three (in either direction) could make a big difference at a single site that happens to be near 1×10^{-4} and would be expensive to remediate. We need a better understanding of the degree of conservatism in both toxicologic and exposure assessments.[9]

Cancer risk assessment has been particularly controversial since its inception. Extrapolation from animals to humans and from high to low doses, as well as the linearity of the dose-response curve at low doses, have always been highly debated assumptions (Graham, Green, and Roberts 1988). Given the powerful influence of cancer potency estimates on reported risks, it is certainly worthwhile to scrutinize their scientific basis, to develop refined estimates for those with initial values, and to devise default values for chemicals that have not yet been adequately tested in long-term rodent bioassays (Taylor, Evans, and McKone 1993).

Because of the potential impact of noncancer assessments on remedial decisions, the methodology for their development also deserves serious scrutiny. The meaning of a hazard quotient or hazard index that exceeds unity is not clear even to the scientific community and therefore is at least equally unclear to many risk managers. Noncancer risk assessment procedures that explicitly indicate the likelihood and severity of adverse health impacts in the exposed population are clearly needed.

Myth 4: **Superfund risk assessments consider only a couple of scenarios and exposure pathways.**

Finding: **Most Superfund site risk assessments evaluate a range of scenarios and exposure pathways.**

Up to this stage in our analysis, we may have contributed to the myth that risk at a Superfund site can be usefully represented by reporting a single, maximum value. While only a few risks reported for each site—often the highest ones—are noted and emphasized in the RODs, the maximum cancer risks and hazard indices provide only a limited picture of the risk profile of a site.

Some critics of the Superfund program have suggested that standard EPA risk assessments at sites are too simplistic to capture the various ways that people might be exposed to contaminants. While there may be some merit to this criticism, we were encouraged by the effort to consider potentially critical exposures that was evident in the risk assessments as summarized in the RODs.

Most risk assessments presented an array of scenarios (such as residential, occupational, or recreational settings) involving different age groups, contaminant concentrations in media on and off the site, and intake routes to characterize potential human exposures. Over 64% used two or more scenarios that may involve several exposure pathways, defined here as unique combinations of environmental media and intake routes (such as ingestion of groundwater). While we were unable to summarize the number of pathways considered at all our sites, most RODs presented results for exposures to two or more media—frequently to one or more intake routes—suggesting the use of several exposure pathways. Groundwater and soil were still by far the most common media discussed. Our findings include only those scenarios considered quantitatively. Many RODs referred to qualitative assessments of other scenarios and pathways that were not recorded in the HCRA database.

The range of cancer risks and hazard indices reported for individual sites is very broad, often spanning many orders of magnitude (that is, factors of ten). About 90% of sites showed differences of at least a factor of ten between the maximum and minimum risks reported and nearly half show differences of five orders of magnitude or more. Keep in mind that the "minimum" values refer to the minimum numbers reported in the RODs and are not necessarily the minimum values for the site as a whole. Indeed, those regions of the site where there is no current contamination and little likelihood of future contamination may not pose any risk.

The variety of scenarios and exposure pathways that come into play at a site, not to mention the huge variability in contaminant concentrations and in potential human activity patterns associated with a site, suggest that risk at Superfund sites is not well represented by a single number. In addition to those factors, the large degree of uncertainty associated with toxicity values and exposure modeling is causing risk assessors to gravitate toward characterizing risks in the form of distributions rather than as single numbers. If Superfund policy is to remain abreast of technical developments in risk assessment, then risk management guidance will need to be revised to reflect the numerical distributions that are being reported to site managers. This conclusion also has important implications for how cleanup goals are ultimately set for sites, in particular for how existing numerical standards and criteria should or should not be used in setting those cleanup goals (see Myth 10).

Myth 5: **The scenario of a child ingesting soil underlies the highest risks at Superfund sites.**

Finding: **Estimated human exposures from groundwater contamination often account for the maximum risks reported at Superfund sites.**

Perhaps the best-known anecdote about Superfund is the scenario of a small child regularly gaining access to and eating small quantities of contaminated soil, a scenario that is presumed to be driving the cleanup efforts at many of our nation's sites. In reality, it is difficult to know what drives a particular decision; a number of factors besides health issues can come into play. However, since public health is the major reason given in most RODs for taking action, we have analyzed the role of the scenario of the child ingesting soil by examining the basis for the highest risk values for each site.

Figures 3 and 4 show the frequency distributions of the maximum cancer risks and hazard indices reported for each site, respectively. In both figures, the left-most bar at each risk level represents all the maximum values, regardless of environmental medium. The maximum risks reported are then separated by environmental medium—groundwater, soil, and all other media (surface water, air, crops, and fish).

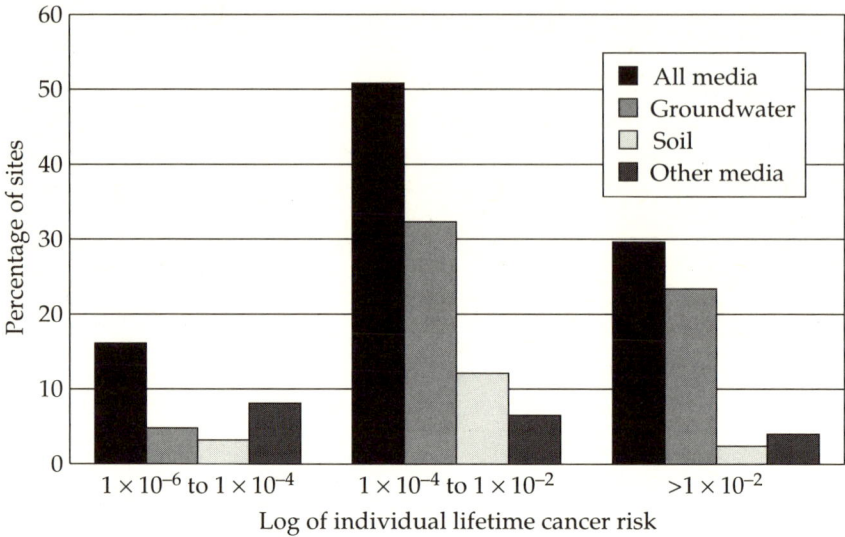

Figure 3. Media Responsible for Cancer Impacts by Level of Maximum Risk per Site

Note: The level of maximum risk per site is based on the maximum individual lifetime cancer risk reported for 124 sites. Each cancer risk is a value specific to a medium or a pathway (that is, does not include combinations of scenarios, media, or intake routes).

Both figures show that exposures to contaminants in soil and groundwater account for most of the maximum cancer risks or hazard index values reported. In only a few cases are exposures to other media important. Exposures to soils, however, account for relatively few of the maximum reported cancer or noncancer risks for the sites in the sample. Only 19% (23) of the 124 sites reporting cancer risks had maximum risks based on soil exposures.

Groundwater exposures account for 62% (77) of the 124 maximum cancer risks and for 63% (67) of 107 of the total noncancer risks. In fact, for cancer risk, the role of groundwater becomes even more dominant at higher levels of risk (greater than 1×10^{-2}).

Of the 23 sites where the reported maximum cancer risk was based on soil exposure, only 13% (3) were based on a child exposed to soil; 78% (18) were based on scenarios involving adult exposures to soil; the remaining two were based on a combined child/adult exposure.[10]

The findings were similar when we examined the basis for the maximum cancer risks reported for all 63 sites at which risks from exposures to soil were quantified. For this analysis, we identified the subset of all sites at which exposures to soil were evaluated and then identified the

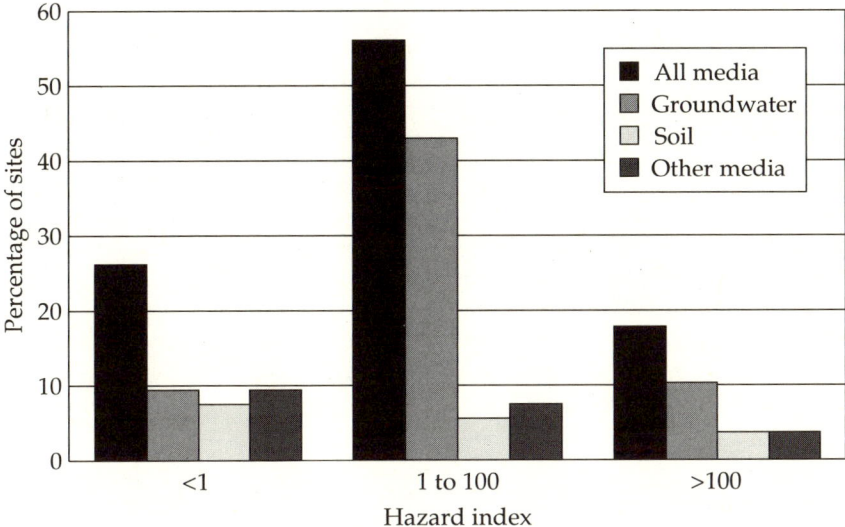

Figure 4. Media Responsible for Noncancer Impacts by Level of Maximum Hazard Index per Site

Note: The maximum hazard index per site is based on the maximum hazard index reported for 107 sites. Each hazard index is a value specific to individual scenarios, media, and intake routes (that is, does not include multiple scenarios, media, or intake routes).

maximum risk for each site in that subset. Although more common, the scenario of the child ingesting soil was still not the most prevalent. Exposures to a child accounted for only 19% (12) of the highest soil risks reported, while exposures to an adult 71% (45) or some combination of adult and childhood exposures (10%) accounted for the remaining soil risks. The percentage of the maximum values accounted for by exposures to a child was again slightly higher for noncancer impacts.

Groundwater contamination, rather than soil ingestion, begins to emerge as a dominant factor in decisions to take action. Assumed exposures—often future exposures—to groundwater most commonly account for the highest risks reported at sites. (About 85% of cancer and noncancer risks associated with groundwater exposure are assumed to take place in the future.) While it is difficult to discern the underlying motives for many decisions from the RODs, the importance of groundwater contamination emerges more fully in our assessment of the basis for site cleanup goals later in this chapter.

Myth 6: **Superfund risk assessments take into account actions that have been taken to limit access to contaminated media.**

Finding: **The estimated health risks at sites are not adjusted to reflect actions that have been taken or might be taken in the future to restrict human access to contaminated materials at the site or to contaminated water supplies.**

Cleaning up contaminated soil and groundwater may seem to be the most direct and complete method of preventing risks to human health, but it is certainly not the only one. The field of public health has a long tradition of using secondary means of prevention, such as purifying water supplies, installing collapsible steering columns in cars to protect drivers who are involved in crashes, and restricting public access to locations or activities that may be hazardous. Similar strategies can be used and are used to protect people from hazardous wastes, but EPA's risk assessment process is not designed to take them into account.

We explored the extent to which access to wastes or to contaminated groundwater and soil had been prevented and the impact that those restrictions had on the risk assessment's assumptions about contact with those media. We found that, at a significant percentage of the sites, alternate water supplies or site access restrictions were being used to limit current exposures, yet EPA assumed eventual exposure via each medium in its risk assessments. The number of sites at which actions have been taken to prevent use of groundwater supplies is probably understated by our database. Several RODs indicated that other types of "institutional" controls (such as deed restrictions specifically to prevent installation of wells) had been put in place; however, our coding of institutional controls does not distinguish among media.

Of the 152 sites in the database, 29% (44) of the RODs indicated that alternate water supplies had been provided. At 57% (25) of these 44 sites, EPA assumed that use of groundwater was occurring or would occur in the future. Even if alternate water supplies are not provided now, most remediation plans call for groundwater monitoring and/or monitoring of drinking water supplies that could be used to determine when alternate water supplies would be needed in the future.

The result for access to contaminated soils or waste on the site was similar. Of the 152 sites in the database, 41% (62) reported site access restrictions, yet exposure to soils was assumed at each of these sites. The restrictions may entail fencing, zoning, or other controls of human access to soil.

The frequency with which exposures are assumed to take place despite the implementation of actions designed to prevent them is not surprising, given EPA guidance on baseline risk assessments. By definition, baseline risks are those that exist or may someday exist in the absence of any remedial action. They are intended to provide the basis for decisions about final remedial actions at the site and, despite the absence of actual exposure, are used to gauge the degree of contamination at the site. The RODs rarely explain, however, why contamination is "hazardous" if it is not expected to result—now or in the future—in human exposure. The "nonuse" value of groundwater may be high or there may be harmful exposures to nonhuman species, but these possibilities are not typically elaborated upon in the RODs. Once again, this deficiency in RODs reflects the poor quality of environmental and welfare analyses that have been conducted at most sites. Without some indication of the likelihood and potential severity of adverse environmental and welfare effects, it is difficult to assess whether extensive soil and groundwater cleanup efforts should be preferred over access restrictions.[11]

Myth 7: **Residential land use assumptions always underlie the highest risks reported for exposures to soil.**

Finding: **Occupational exposure scenarios account for a large fraction of the highest soil risks reported.**

A corollary of the myth that risks associated with small children eating soil motivate cleanup decisions at most Superfund sites is the notion that residential land use assumptions are an important factor in the high risks estimated for Superfund sites. Much recent debate has focused on the selection of appropriate land uses on which to base exposure assessments. The concern is that unrealistic assumptions about existing and future use of a property (such as siting a residential development on a landfill) lead to an overstatement of potential risks. While we have not yet fully addressed the impact of land use on all risk estimates, we have assessed its impact on the maximum soil risks reported in our survey.

Questions about EPA's assumptions about the ultimate uses for Superfund sites are warranted. For example, EPA does assume that residents will occupy industrial sites at some point in time, whether those sites currently house active facilities or not, for about half the industrial sites in our survey. The impact of these types of assumptions on risk estimates, however, is not uniform. While the high groundwater risks reported for the sites in our survey are almost exclusively based on assumptions about residential use of groundwater supplies, the maximum reported soil risks are not.

If one examines the underlying scenarios for the highest soil risks reported, occupational scenarios account for a surprisingly large fraction. Occupational scenarios are ones in which an adult worker is assumed to ingest, inhale, or have dermal contact with soil or dust inadvertently while working on the site.[12] As discussed in Myth 5, soil exposures accounted for the maximum risk reported at only 23 sites. Most of these risks assumed exposure to adults; half were grounded in occupational scenarios.

The results were similar for the maximum soil risks at the 63 sites that reported soil risks of any kind. Forty-five (71%) maximum soil risks were due to adult exposures; nearly all were based on occupational scenarios. When all age groups were taken into account (child, adult, and child/adult categories), occupational scenarios accounted for about one third of the maximum risks reported, with residential and, to a lesser extent, recreational scenarios making up most of the remainder.

These findings concerning the importance of occupational exposure scenarios have possible implications for current proposals to modify EPA's land use policies. Restricting the use of residential development as the default land use in favor of more "realistic" and perhaps more lenient land use assumptions involving industrial use of the site may not always lead to less cleanup. The assumptions underlying the occupational scenario have not received as much scrutiny as those scenarios that have been suspected of "driving" cleanup (that is, residential groundwater and soil ingestion). This somewhat unexpected finding from our survey underscores a need for a closer look at the precise nature of occupational exposures that might occur with and without remediation of a site.

Myth 8: **Superfund sites are contaminated with many chemicals, all of which contribute significantly to public health risk.**

Finding: **A small number of chemicals account for the risks reported at many Superfund sites.**

In contrast to the hundreds of chemicals for which media at Superfund sites are routinely analyzed and the large numbers that are often detected, a relatively small number of contaminants typically contribute substantially to risk at a site. At the majority of sites, one or two contaminants together accounted for most of the risk. The results were qualitatively similar for noncancer hazard index values, with a somewhat larger number of values dominated by three or more contaminants.

To evaluate this issue, we identified the contaminants that together contributed 95% of the individual cancer risk or hazard index value. In the absence of quantitative information, we listed the contaminants identified by the ROD as the most important contributors to health risk. These contaminants, identified using either quantitative or qualitative approaches, were defined as "driving" the risk or hazard index values. Our analysis of this issue was potentially compromised by the fact that contaminant-specific information accompanied only 775 of the 1,335 risk or hazard index values recorded in our database. However, the percentages of contaminants contributing to risk and the identities of the key contaminants have remained essentially the same throughout the collection of data.

We examined which contaminants were driving the maximum reported cancer and noncancer numbers. Figure 5 indicates the ten contaminants most frequently cited as contributing to maximum cancer risk. Although these contaminants were generally dominant sources of risk, at specific sites other contaminants may be critical. Among the sites we evaluated, about forty contaminants were cited at least once.

Among the carcinogenic contaminants cited, arsenic topped the list, followed by a series of volatile organic compounds. The list for noncarcinogens differs somewhat from that for carcinogens, with more inorganic compounds, chromium, lead, manganese, nickel, and barium joining the list. Arsenic, 1,1-dichloroethene, and tetrachloroethene were the most frequently reported contaminants for both cancer and noncancer risks.

The relatively small number of contaminants contributing to the cancer and noncancer risk values reported may reflect the fact that many chemicals detected at sites may not have undergone toxicologic review or have been assigned toxicity values, whether a cancer slope factor or an RfD, as reported in EPA's Integrated Risk Information System (IRIS). IRIS is the primary source for all toxicity values used in Superfund risk assessments. The impact of excluding the sometimes large number of contaminants with no toxicity values has not been assessed but certainly understates risks at these sites to some degree.

At the same time, the small number of contaminants that determine the quantitative risks reported and that are most commonly identified underscore the importance of careful assessment of their toxicity. The toxicologic assessments of the carcinogenicity and noncancer health impacts for these contaminants factor heavily in the risks reported at many sites and ultimately in the remedies and goals selected. Since IRIS toxicity values play a crucial role in Superfund risk assessment, it is critical that these values are based on the best available scientific data and judgment.

Careful assessment of the possible influence of background levels of contaminants on reported risk is also critical, as the ubiquitous presence of arsenic suggests. While arsenic can be associated with disposal activities at a site, it is also naturally present in many soils at levels that would be associated with risks that fall within or exceed EPA's allowable risk range. Few of the RODs reviewed discussed background levels of contaminants, so it was not possible to determine the sources of the arsenic so frequently assessed. Given their potential importance in determining

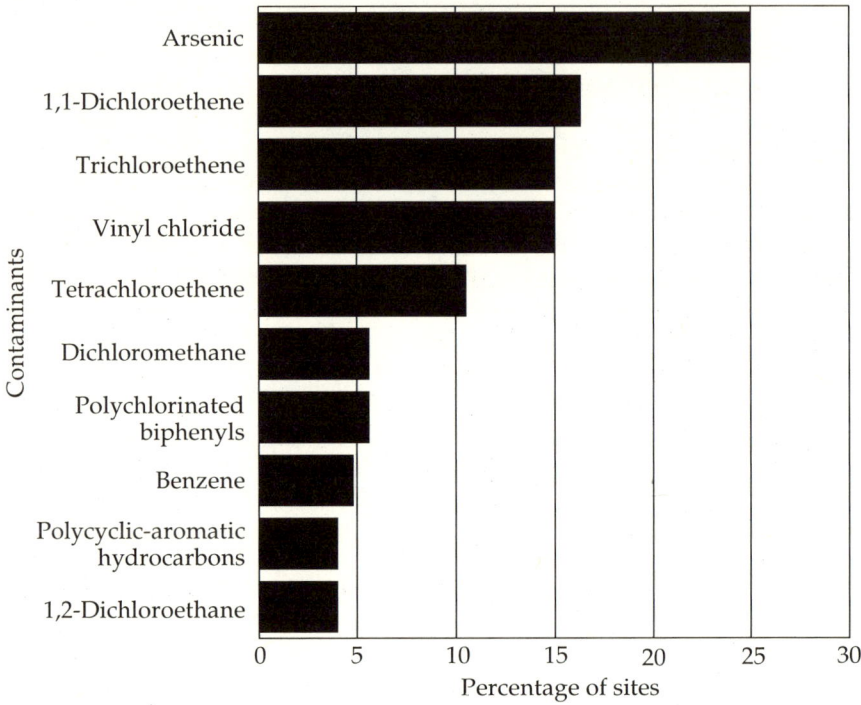

Figure 5. Top Ten Contributors to Cancer Risk

Note: This frequency distribution is based on the contaminants that accounted for the maximum cancer risks reported at 124 sites. Cancer risks were not reported at all sites.

risk levels and consequently the need for remedial action, more explicit consideration of the role and policy implications of background levels of contaminants is needed.

Myth 9: **Remedial decisions take into account the number of people potentially adversely affected at Superfund sites.**

Finding: **Remedial decisions at sites are typically made without even rough information about the potential size of the exposed population, the number of expected excess cancers, or the number of citizens with increased chances of suffering noncancer health effects.**

EPA's use of information about the size of the exposed population and population risks to inform decisionmaking varies across its regulatory programs. In no situation has Congress prohibited risk assessors from presenting to risk managers information about how widespread a pollution problem may be (Rosenthal, Gray, and Graham 1992). One therefore might reasonably expect that Superfund risk assessments would provide site managers with information about how many people might be adversely affected—now and in the future—by a contaminated site. In fact, estimates of population risk have little place in the Superfund program.

Few RODs give any sense of the number of people near the site; only 24% (37) of the sites in the current sample gave even the population of the nearest town. A smaller fraction (11%) gave at least some estimate of the population within one mile of the site. While such estimates are only crude indicators of the likely exposed population, they begin to place the issue of exposure into some perspective.

Population estimates that better characterize the size and nature of the populations potentially exposed under the various exposure scenarios would be more informative. However, such discussions are rare in RODs; of the 1,335 risk estimates reported in the database, only 6% provided any characterization of the populations potentially exposed. Only one site attempted to estimate an overall population risk: the number of individuals who would be expected to develop cancer under the circumstances envisioned in particular scenarios and exposure pathways. The lack of emphasis on populations at risk, however, has undermined the collection of data necessary to evaluate the extent of actual exposures or to more accurately project exposures over time at individual sites.

The risk management criteria employed by EPA at Superfund sites are questionable in the absence of information about the number of people at risk. A cancer risk level of 1 in 10,000 or less may reasonably be regarded as negligible if only a handful of people are exposed to this level of risk, but the matter is entirely different if over one million people may be exposed. Furthermore, current Superfund risk assessments, by omitting information on population risk, do not provide Congress the information and perspective it needs to assess the relative importance of the public health risks posed by Superfund sites compared to other pressing environmental and public health issues. EPA's Risk Assessment Council issued guidance in November 1991 aimed at promoting more complete characterizations of risk throughout the agency, but it is not clear if this guidance has actually changed analytical practices in the Superfund program (U.S. EPA 1991a).

Myth 10: **The concept of acceptable or negligible risk underlies the development of cleanup goals at Superfund sites.**

Finding: **Despite the substantial effort invested in site-specific risk assessments, they are playing a relatively minor role in defining cleanup goals.**

Standard risk assessment procedures were established by EPA as a way of evaluating the potential health, welfare, and environmental impacts of hazardous wastes. Procedures were needed because of the scarcity of existing standards or criteria by which to judge when problems might result from the many contaminants identified at Superfund sites. In light of the motivation for Superfund risk assessment, one might reasonably expect that the process of setting cleanup goals would also be informed by a site's risk assessment. In turns out that this is also a myth. In the RODs reviewed for this study, site-specific risk assessments played little role in the establishment of cleanup goals.

Not all of the 152 RODs for which remedial action was taken set explicit cleanup goals, although they specified such goals more often for groundwater than for soil. A total of 63% (95) of RODs stated numerical goals for groundwater but only 38% (58 sites) stated goals for soils. Very few RODs set goals for media other than groundwater and soil; since this finding is consistent with the relatively low risks reported for other media, their goals are not discussed further here. The reason that so many RODs omitted specific goals was not always clear. A few RODs did indicate that meeting the allowable "risk range" was the goal.

If achieving acceptable or negligible risk given the exposures evaluated in the site-specific risk assessment is not the goal, then what is? It turns out that the environmental goals specified in other local, state, and federal programs are often used by site managers as the operative cleanup goals. For example, a federal drinking water standard may be used to define cleanup targets for groundwater. These existing federal standards or criteria, whether "applicable or relevant and appropriate requirements" (ARARs) or "to be considered" (TBCs),[13] are the most frequently cited cleanup goals. The concept of ARARs was established under the 1986 amendments to the Superfund law (Superfund Amendments and Reauthorization Act, or SARA) to promote consistency among the various state and federal regulations. The importance of ARARs in setting goals partly reflects the requirement in SARA that they be met by any remedy selected.

In the case of groundwater, ARARs dominated the selection of cleanup goals at Superfund sites. A total of 98% of the 95 sites with goals for groundwater relied on ARARs alone or in combination with the development of goals based on site-specific risk assessments, while 63% (60) relied on ARARs alone. Only two sites relied on risk-based goals alone.

Since there are many laws and regulations that define "safe" levels of contaminants, it is useful to consider which ones are applied most frequently at Superfund sites. As Table 1 shows, the most frequently used ARARs are the Maximum Contaminant Levels and Maximum Contaminant Level Goals (proposed or final) under the Safe Drinking

Water Act, and state drinking water or groundwater standards or criteria. Among the several RODs employing state standards or criteria for cleanup goals were four RODs specifying Pennsylvania's requirement that groundwater be remediated to background levels found in nature. Surface (or "ambient") water quality criteria, developed for rivers, lakes, and streams under the terms of Water Pollution Control Act, were also cited in a number of instances.

ARARs were less frequently cited as the basis for soil goals than for groundwater, reflecting the current absence of any federal standards for contaminants in soil. Of the 58 sites with goals for soil, 52% (30) relied on ARARs alone or in some combination with risk-based goals; 30% relied on ARARs alone. Thirty-one percent (18) of the sites, compared to 2% (2) in the discussion of groundwater goals, used risk-based goals exclusively.

Protection of groundwater emerges as a critical factor in the development of cleanup standards for soil (see Table 1). While state soil standards or criteria formed the basis for about one quarter of the ARARs reported, drinking water standards or other standards to protect groundwater underlay a still larger fraction (40%). However, the overall percentage of soil cleanup goals based on groundwater protection might be higher than this figure suggests. Some state soil standards—Resource Conservation and Recovery Act regulations as well as risk-based goals—are often set at levels that protect groundwater from the potential leaching of contaminants from soil.

Table 1. Applicable or Relevant and Appropriate Requirements (ARARs) Forming the Basis for Groundwater and Soil Goals

	Groundwater[a] (percent)	Soil[b] (percent)
Maximum Contaminant Level (MCL)	38	22
Maximum Contaminant Level Goal (MCLG)	15	8
Proposed MCL/MCLG	13	5
State drinking water or groundwater standards or criteria	14	5
Health advisories	2	
Ambient water quality criteria	9	
State soil standards or criteria	N.A.[c]	24
RCRA land disposal regulations	N.A.	8
Other	9	27

[a]Based on 125 cases in which an ARAR was specifically reported in a ROD as a basis for a groundwater goal. Individual RODs may list more than one ARAR as the basis for its cleanup goals, so this number is higher than the 95 sites reporting any numerical goal for groundwater.
[b]Based on 37 cases in which an ARAR was specifically reported in a ROD as a basis for a soil goal. Percents do not sum to 100 because of rounding.
[c]N.A. = not applicable

The distinction we are making here between ARARs and "risk-based" goals, although commonly made in the Superfund debate, is somewhat contrived because ARARs themselves are often rooted in concepts of acceptable or negligible risk. For example, many chemical-specific ARARs are based on numerical calculations that employ cancer slope factors, toxicity values, and specific exposure assumptions. For groundwater cleanup goals in particular, the underlying exposure assumptions used in risk-based goals are usually the same as those used to develop ARARs, such as the assumptions underlying the Maximum Contaminant Level Goals (MCLGs) under the Safe Drinking Water Act. ARARs, it should be noted, were typically developed without thought given to their application to Superfund sites, where multiple contaminants are typically present. The fundamental issue is that the rationale for many ARARs has not been subjected to rigorous policy analysis, which raises further questions about whether they should be exported directly into the Superfund program.

Thus, important factors—such as the size of the potentially exposed population, the feasibility of meeting goals, and the cost-effectiveness or even affordability of such goals—are not typically considered when ARARs are established, and such factors are certainly not considered when they are applied to specific Superfund sites.

The goals set under the Safe Drinking Water Act and their application to the Superfund program illustrate some of the hidden policy judgments. The Safe Drinking Water Act calls for EPA to establish both MCLGs, which are intended to be fully protective of public health, and Maximum Contaminant Levels (MCLs), which are the legally binding maximums that may take into account feasibility considerations. For cancer-causing chemicals, the MCLGs are often set at zero because EPA assumes that any exposure to a cancer-causing chemical, no matter how small, causes an increase in cancer risk. Since zero contamination is not practical for regulatory purposes, EPA may then set the MCL for a carcinogen at the lowest level that can be detected with current methods of analytic chemistry or (if that is still too stringent to be practical) at a more permissive level that is achievable and affordable for municipalities. MCLs are not always as protective as the "allowable risk range" that is prominently cited in the National Contingency Plan for use at Superfund sites. Since most sites have more than one contaminant of concern, many of which do not have MCLs, MCLGs, or other standards, the total residual risk for the site may exceed the allowable risk range by a considerable margin.

ARARs are being used so extensively in the Superfund program that one can reasonably question the actual role of site-specific risk assessments. After all, ARARs are already used in conjunction with the risk assessment to decide whether the site needs to be remediated, with the presumptions being that remediation is necessary if the ARARs are exceeded and ARARs are the targets that remedial actions are aimed at achieving. Congress needs to choose: either make remedial decisions and set goals on the basis of risk-based policy, including a site-specific risk assessment, or on the basis of ARARs. Sound principles of risk management would not seem to favor such widespread use of ARARs because they were established without any sensitivity to site-specific conditions, population density, feasibility, or affordability. Some form of risk-based policy, which includes administrative flexibility to respond to unique features of sites, is a logical direction for the future.

IMPLICATIONS FOR SUPERFUND REFORM

Before discussing our opinions on the implications of our findings to possible reforms in the Superfund law, it is worth restating the findings discussed in this chapter:

- Public health is the primary rationale given for decisions to take remedial action at Superfund sites.
- Both the estimated noncancer risks and cancer risks typically meet EPA criteria for taking remedial action.
- Neither increasing nor decreasing the degree of conservatism alone is likely to profoundly influence remedial decisions.
- Most Superfund site risk assessments evaluate a range of scenarios and exposure pathways.
- Estimated human exposures from groundwater contamination often account for the maximum risks reported at Superfund sites.
- The estimated health risks at sites are not adjusted to reflect actions that have been taken or might be taken in the future to restrict human access to contaminated materials at the site or to contaminated water supplies.
- Occupational rather than residential scenarios account for a large fraction of the highest soil risks reported.
- A small number of chemicals account for the risks reported at many Superfund sites.
- Remedial decisions at sites are typically made without even rough information about the potential size of the exposed population, the number of expected excess cancers, or the number of citizens with increased chances of suffering noncancer health effects.
- Despite the substantial effort invested in site-specific risk assessments, they are playing a relatively minor role in defining cleanup goals.

Many of the myths that have surrounded the Superfund program relate to the health and environmental risks associated with hazardous waste sites, in particular the magnitude of those risks, their theoretical and technical underpinnings, and their role in decisionmaking. Using analyses of a comprehensive database of 1991 RODs, HCRA has dispelled many of the common myths or misperceptions about risks in the Superfund program. At the same time, our findings indicate that legitimate concerns about the conduct and role of risk assessment remain. Questions about the characterization of potential noncancer and environmental impacts, the role of population risk, and the appropriateness of land use and other exposure assumptions, among others, all merit serious thought.

These questions regarding the process of risk assessment, however, are secondary to the issue of the proper role of risk assessment in the Superfund program—an issue Congress must face in the reauthorization of Superfund. It is an issue that is complicated by the tension between the more explicit, visible goal of protecting public health and the more understated, implicit objective of protecting groundwater resources.

As the program is currently being implemented, risk assessment faces an increasingly minor role in remedial decisions. For example, although the baseline human health risk assessment is used to some extent to determine if remedial action is necessary, this same determination is often made by comparing the degree of soil and groundwater contamination to ARARs, as noted in Myth 10. When ARARs are not satisfied, remedial action is often judged to be necessary. In any event, most of the sites in our survey of 1991 RODs have sufficiently serious amounts of contamination in at least some portion of the site that taking no action is not a possibility, whether the risk management criterion is compliance with ARARs or conformance with the allowable cancer risk range and hazard index value.

Superfund risk assessments play even less of a role in defining cleanup targets at sites, since it is ARARs that seem to provide the binding basis for contaminant-specific cleanup goals. This finding is troubling because most ARARs have a weak policy rationale to begin with and were certainly not intended to be applied to Superfund site cleanups when they were devised by federal and state agencies. Since it is the stringency of the cleanup goal, in conjunction with the type of remedy, that presumably determines the cost of remedial activities, it is disturbing that the cleanup goals are not set based on site-specific risk assessments and management considerations. For example, salient policy considerations (such as the size of the exposed population, the natural background levels of contaminants, and the costs of achieving various degrees of risk reduction) are not considered when cleanup goals are set at Superfund sites.

Congress and EPA are faced with a clear choice: either they should eliminate the minor role that risk assessment currently plays in the program and thereby save the time and cost to all parties associated with such assessments, or they should strengthen the role of risk assessment in remedial decisions by adopting more flexible, site-specific risk management criteria and requiring risk assessments that utilize the best available science in considering the full range of public health, environmental, and welfare effects. Little rationale exists for retaining the current practice of producing complex risk assessments that minimally affect remedial decisions. Given this choice, we would favor a stronger risk assessment because we believe that Congress and the public will demand an accounting of what is being accomplished at Superfund sites for the billions of dollars that will be spent. Only a concerted effort at enhancing the processes of risk assessment and management can produce the kinds of information necessary to rebuild the credibility of this troubled program.

If Congress and EPA should decide to strengthen the role of site-specific risk assessments in the forthcoming reauthorization of Superfund, it is critical that some of the defects in the current risk assessment process be corrected. Correcting these defects will improve the analytical process in the future.

First and most importantly, Congress needs to insist that EPA investigate the environmental and welfare risks that sites pose in addition to the current and future health risks. Within the concept of welfare effects, we include the nonuse value of groundwater, which includes the psychological comfort of knowing that groundwater is clean. In the case of health risks, EPA needs not only to refine cancer risk assessment procedures as science progresses, but also to develop new methods of noncancer risk assessment that indicate the potential frequency and severity of effects that may occur within exposed populations. Only with the benefit of this more comprehensive assessment of site-specific risks can policymakers make informed decisions about how much of our nation's scarce resources should be devoted to cleaning up hazardous waste sites.

Second, EPA needs to go beyond estimating the health risk to hypothetical individuals living near sites and to produce distributional estimates of risk that convey the potential size of exposed populations (now and in the future), including the fate of particularly susceptible and/or highly exposed subpopulations. Superfund policy will not remain abreast of technical developments in risk assessment unless EPA periodically publishes guidance on how distributional methods of uncertainty and variability analysis should be applied in site-specific assessments.

Finally, human health risk assessments need to be conducted with realistic assumptions about future land use, public access to sites, and availability of alternative water supplies when groundwater is contaminated. Perhaps one of the unexpected benefits of more realistic assumptions in human health risk assessment will be an explicit public debate about the potential environmental and welfare effects of not cleaning up sites.

One of the greatest myths about the current operation of the Superfund program, which is perpetuated by EPA's official rationales for remedial decisions, is that remediation is justified by the need to protect people's health from consumption of contaminated groundwater. For if that were the primary rationale for action, there are many feasible options short of treating contaminated groundwater (such as providing alternative water supplies) that could protect public health. Our analysis of the Superfund program suggests that one of the hidden yet worthy objectives of the program is to protect the quality of our nation's groundwater for future yet unspecified uses by humans and nonhuman species. The question that policymakers need to address, but cannot address in the face of Superfund mythology, is how much as a nation are we willing to pay now to protect the cleanliness of groundwater supplies for unspecified future uses and to have the comfort of knowing that the groundwater is clean. Reasonable people will disagree about the answer to this question, but the question needs to be asked.

ACKNOWLEDGMENTS

The authors, who are researchers at the Harvard Center for Risk Analysis, Harvard School of Public Health, gratefully acknowledge the support and thoughtful input from John Evans, Kim Thompson, Linda Greer, Sarah Levinson, Doug Sarno, Michael Parr, and William Shively throughout the development of this project. A special thanks goes to Adrienne LaPierre for her substantial contributions to the early pilot of the study. Finally, we are indebted to Mia Costic, Elizabeth Coughlin, Cheryl Keenan, Josh Stebbins, Lise Wilson, Cliff Wood, and Hui Zhou for their faithful and painstaking review of the Records of Decision. The views expressed in this chapter are those of the authors.

ENDNOTES

[1]Doty and Travis (1989) surveyed approximately 50 sites with 1987 RODs. A later study of 110 wood treatment and PCB (polychlorinated biphenyl) sites

was undertaken by Gupta, Van Houtven, and Cropper (1993). Viscusi and Hamilton have also recently developed a database of sites with RODs in 1991; see Chapter 3 in this book for details of their work.

[2]While numerous databases at EPA are used to track various aspects of the Superfund program, they differ widely in their scope, structure, software, completeness, and quality, as well as their accessibility to independent researchers. The development of a new comprehensive database, 3 DB, to assist in policy analysis of the Superfund Program is underway at EPA and is expected to be complete in spring of 1995.

[3]Cancer risk is reported as a hypothetical individual's lifetime excess risk of cancer due to exposures originating from the site. It is calculated as the product of dose expressed in units of milligrams (mg) of the chemical per kilogram (kg) of body weight per day (mg/kg/day) and a "slope factor" (sometimes called a potency factor) in inverse units of mg per kg per day or $(mg/kg/day)^{-1}$.

This product of the dose and slope factor is a unitless number between 0 and 1, a probability that typically is expressed in exponential form. For example, a cancer risk expressed as 1×10^{-6} represents 1 chance in a million of contracting cancer. For the average individual, whose actuarial risk of developing some form of fatal cancer in his or her lifetime is 1 in 4 or 0.25, an increased risk of 1×10^{-6} results in a total risk of 0.250001. The equivalent expression of this risk level as a "population risk" is that if one million people were exposed to the level of pollution associated with a 1×10^{-6} risk, one person would be expected to develop cancer.

[4]Noncancer assessments are summarized with a simple ratio: the estimated daily dose of a chemical, again expressed in mg/kg/day, divided by the *reference dose* (RfD) for that chemical—a number that was once called an "acceptable daily intake." The RfD for each chemical is developed by EPA typically by dividing either the No Observed Adverse Effect Level (NOAEL) or the Lowest Observed Adverse Effect Level (LOAEL) from animal tests by a series of safety factors (usually one or more factors of 10). The NOAEL or LOAEL is used to approximate the threshold dose, while the safety factors are applied to account for concerns about the quality of the underlying toxicologic data, the relative sensitivity of humans and animals, and other issues. The ratio of the estimated intake to the RfD for a particular chemical is called the *hazard quotient*. When hazard quotients are summed for several chemicals and/or, in some cases, for multiple scenarios or pathways of exposure, the resultant value is called the *hazard index*.

[5]While many Superfund sites have more than one area that requires remedial action, often defined as "operable units" by EPA, most of the RODs currently in the database contain information relating to the whole site. Only about 6% of sites reported more than one operable unit.

[6]All of the RODs contain standard legal language asserting that the selected remedy protects public health, welfare, and the environment. In their discussions, most RODs emphasize or highlight particular risks to back up that assertion while sometimes discounting others. Such highlighted risks are those the HCRA database asked the ROD reviewer to identify. Although such a question may appear perilously subjective, readers reviewing the same RODs have been in good agreement.

[7]While the use of maximum risk is an oversimplification of the risk assessments conducted for sites, it represents an attempt to represent each site and exposure pathway only once in the analysis. As there is substantial variability among sites with respect to the extensiveness of the risk assessment conducted, a simple analysis of all risks reported for all sites would likely result in the overrepresentation of some sites and pathways, and thereby introduce a fundamental source of bias into the results.

[8]One possible explanation for the unexpected magnitude of the maximum hazard indices is the manner in which the HCRA database records the hazard index. The database reports calculated the hazard index as the sum of hazard quotients for all contaminants for a particular exposure pathway. Current EPA guidance (U.S. EPA 1989) suggests that hazard indices may be reported for subsets of contaminants segregated by target organ. Use of hazard indices for these subsets would tend to result in reporting lower maximum hazard indices. However, few of the sites included in this analysis presented hazard index by target organ.

[9]An important issue in the debate about the possible degrees of overstating risk is the plausibility of the scenarios: not just of the individual assumptions about the frequency and duration of exposure but of the scenario itself. How likely is a scenario to represent existing or future activities on or near a particular site? We have not addressed this form of uncertainty at this time but it remains an important concern.

[10]The breakdown appeared to be slightly different for noncancer risks. Although not the majority, more of the maximum hazard index values were based on children's exposures to contaminated soil. This finding makes sense given the different ways in which dose is calculated for cancer and noncancer risk estimation. For cancer risk assessment, the dose is averaged over a lifetime whereas for noncancer risk assessment, the dose is simply the average daily dose during the period of exposure. Consequently, if a child's dose during a five-year period (say, from ages one to six years) is averaged over a seventy-year lifetime, the average daily dose is much lower than the dose during that five-year period of exposure.

[11]The need for a better understanding of potential welfare and environmental effects and the value we place on them as a society is underscored by the expense of more permanent remedies. When viewed from the perspective of public health alone, the expense may sometimes seem difficult to justify. At the same time, the simplicity of access restrictions and alternate water supplies may be deceiving: such actions must be maintained over long periods of time, sometimes hundreds of years, and can therefore themselves be expensive. More careful analysis of both these issues is critical.

[12]The occupational scenario is distinct from another worker exposure scenario, typically called a *construction scenario,* in which workers are exposed while excavating and/or building on the site.

[13]We have included in an ARARs category both requirements that strictly may be considered ARARs and those that may be more correctly characterized

"to be considered" (TBCs), since the RODs frequently did not distinguish between the two and did not make it clear how they would be treated differently in the implementation of the remedy.

REFERENCES

Doty, C., and C. Travis. 1989. The Superfund Remedial Action Process. *Journal of the Air Pollution Control Association* 39 (12): 1,535–1,542.

Finkel, A.M. 1989. Is Risk Assessment Really Too Conservative? Revising the Revisionists. *Columbia Journal of Environmental Law* 14: 427–467.

Graham, J.D., L.C. Green, and M.J. Roberts. 1988. *In Search of Safety.* Cambridge, Mass.: Harvard University Press.

Gupta, S., G. Van Houtven, and M. Cropper. 1993. Cleanup Decisions under Superfund: Do Benefits and Costs Matter? *Resources* (Spring) 111: 13–17. Washington, D.C.: Resources for the Future.

Hazardous Waste Cleanup Project. 1993. *Exaggerating Risk: How EPA's Risk Assessment Distorts the Facts at Superfund Sites throughout the United States.* Washington, D.C.: Morgan, Lewis and Bockius.

Hegner, R.E. 1994. Does Protecting for Human Health Protect Ecological Health? *Risk Analysis* 14 (1): 3–4.

Rosenthal, A.A., G.M. Gray, and J.D. Graham. 1992. Legislating Acceptable Cancer Risk from Exposure to Toxic Chemicals. *Ecology Law Quarterly* 19: 269–362.

Taylor, A.C., J.S. Evans, and T.S. McKone. 1993. The Value of Animal Test Information in Environmental Control Decisions. *Risk Analysis* 13 (4): 403–412.

U.S. EPA (Environmental Protection Agency). 1989. *Risk Assessment Guidance for Superfund: Volume I, Human Health Evaluation Manual (Part A), Interim Final* (December). Washington, D.C.: U.S. EPA.

———. 1991a. *Guidance for Risk Assessment.* Washington, D.C.: U.S. EPA, Risk Assessment Council.

———. 1991b. *Role of the Baseline Risk Assessment in Superfund Remedy Decisions,* OSWER Directive 9355.0–30. Washington, D.C.: U.S. EPA, Office of Solid Waste and Emergency Response (OSWER).

3

The Magnitude and Policy Implications of Health Risks from Hazardous Waste Sites

James T. Hamilton and W. Kip Viscusi

The impetus for the Superfund program was a concern with the impact of hazardous chemical waste on human health. The risks posed by exposure to contaminants from Superfund sites generated alarm among residents surrounding these sites, stimulating Congress to draft cleanup legislation and regulators to design site remediation procedures that imposed costs on potentially responsible parties. The U.S. Environmental Protection Agency (EPA) now requires a risk assessment as part of the process of analyzing remediation options at sites on the National Priorities List (NPL), which is the set of sites that EPA has designated as being eligible for remediation funds.

A substantial amount of information exists in the administrative record for each NPL site on the quantitative assessment of cancer and noncancer risks for different populations potentially affected by chemicals from the site. In addition, the data produced by actions and studies at Superfund sites have generated debates over the progress and stringency of cleanups (Doty and Travis 1989; Hird 1990), the assumptions employed in the exposure scenarios analyzed (VERSAR 1991; Burmaster and Harris 1993; ENVIRON 1993; Hazardous Waste Cleanup Project 1993; U.S. EPA 1993), and the costs of reducing these risks (Russell, Colglazier, and English 1991; Acton and Dixon 1992; Gupta, Van Houtven, and Cropper 1993).

To date, however, researchers have not systematically explored the data on human health risks at Superfund sites, in part because of the significant transaction costs associated with collecting this information from EPA regional offices. This gap in our knowledge of Superfund risks is particularly striking given the current policy context. The Superfund program is scheduled for reauthorization in 1994,[1] and Congress must decide whether to continue the effort in its current form or modify it in some manner. Clearly, no sensible decision can be made with respect to reauthorization unless one understands the program's fundamental mission, the reduction of risks from hazardous wastes. To date, there has been little basis for making informed judgments other than anecdotal evidence, which has generated inconsistent portrayals of the risks. Some case studies highlight the salience of hazardous wastes and their importance to affected residents, while other accounts attempt to trivialize the risks by claiming that the sites are only dangerous to children who eat large quantities of contaminated dirt.

IDENTIFYING THE RISKS: AN OVERVIEW OF OUR ANALYSIS

Our analysis represents the first systematic effort to document the character of the risks addressed by Superfund. Which population groups are most affected? How do these risks arise? What is the magnitude of the risks that are present? Are the risks in fact trivial, or are there serious threats to public health?

Most important from a policy standpoint, should we expand the range of policy options being considered? For instance, should we continue to favor the more "permanent" options of hazard treatment and removal over on-site containment and land-use restrictions? Although our analysis does not assess all attributes of hazardous waste policies, it does highlight a key dimension—the pathway by which the risks arise. Limited policy options, such as capping and fencing a site or restricting land use, can eliminate some mechanisms by which risks arise. Our analysis documents the nature of these risk pathways and provides startling evidence on the way in which these risks arise.

The character of the risks is in many respects quite surprising. Whereas risks to current residents have played a pivotal role in generating political support for the Superfund program, *the overwhelming preponderance of the risks is to future populations* for land uses that represent departures from current behavior. In our database of seventy-eight sites, there are thirty-five sites where future residential pathway risks occur despite the absence of any current residential risks exceeding the 10^{-6} cutoff level used in this analysis. The detailed analysis of the character of the risks, which leads to a wide variety of similar insights, is particularly instructive in highlighting how different policy mechanisms can influence the pathways responsible for generating the risk.

The Scope of Our Analysis

To address these issues, we analyze the human health risk assessments conducted at seventy-eight Superfund sites that had Records of Decision (RODs) signed in 1991 or 1992. We focus on the distribution of these risks across different categories of analysis in risk assessments. These categories of analysis include:
- time frame of exposure (current use or potential future uses)
- location of exposure (on-site or off-site)
- population type (residential, worker, recreational, trespasser)
- exposure medium (such as soil or groundwater)
- exposure route (such as ingestion, dermal contact, inhalation)

This chapter does not link these risk assessments to population data to derive estimates of potential deaths or diseases, conduct sensitivity analysis of the assumptions used in the risk assessments, or derive estimates of the costs of dealing with these risks. These issues will be explored in future research with an expanded version of this data set. This chapter focuses instead on detailing the types of cancer and noncancer risk data used by regional EPA officials to reach the remedy selections contained in the RODs.

In the rest of this chapter we describe our analysis, focusing on the following topics in the order noted:
- the risk assessment data developed at Superfund sites, presenting the legislative and regulatory context and discussing both the evaluation of risks to human health and the risk assessment pathways
- the construction of our database
- the results of our analysis of the cancer and noncancer risk pathway data from these seventy-eight sites
- the magnitude of the risks associated with different pathways
- a breakdown of the chemicals involved with these pathways
- conclusions about the lessons for policy analysis from these results

RISK ASSESSMENTS AT SUPERFUND SITES

In order to place this analysis in context, it may be instructive to first understand the historical and regulatory context out of which Superfund emerged, and then to illustrate the concepts of the baseline risk assessment and risk pathways.

Legislative and Regulatory Context

Congress passed the Comprehensive Environmental Response, Compensation, and Liability Act (CERCLA) in 1980, amid a debate that focused on the potential health hazards posed by leaking chemicals at sites such as Love Canal. Section 104(a) of this act, which created the Superfund program, directed EPA to respond to hazardous waste sites that pose "a substantial endangerment to public health, welfare, and the environment." In the Superfund Amendments and Reauthorization Act (SARA) passed in 1986, Congress stated that remedial actions were preferable if they employed treatment methods that permanently and significantly reduced the volume, toxicity, or mobility of hazardous substances at sites (Section 121). The law also required that Superfund remedial actions comply with federal standards considered to be "applicable or relevant and appropriate requirements" (ARARs) and that state ARARs be met at Superfund sites if they are stricter than federal ones.

The regulations implementing the Superfund law, the National Contingency Plan (55 *Federal Register* 8665–8865, March 8, 1990), require that a site-specific baseline risk assessment be conducted to "characterize the current and potential threats to human health and the environment that may be posed by contaminants migrating to ground water or surface water, releasing to air, leaching through soil, remaining in the soil, and bioaccumulating in the food chain" (Section 33.430(d)(4)). In a directive published in April 1991 on the role of the baseline risk assessment in the Superfund remedy selection, the Office of Solid Waste and Emergency Response stated that "where the cumulative carcinogenic site risk to an individual based on reasonable maximum exposure for both current and future land use is less than 10^{-4} and the noncarcinogenic hazard quotient is less than 1, action generally is not warranted unless there are adverse environmental impacts" (U.S. EPA 1991). Regional EPA decisionmakers may choose to take action at sites with cancer risks smaller than 10^{-4}, but RODs with remedial actions at sites with risks "within the 10^{-4} and 10^{-6} range must explain why remedial action is warranted." The directive also declared that the "EPA uses the general 10^{-4} to 10^{-6} risk range as a 'target range' within which the Agency strives to manage risks as part of a Superfund cleanup." This means that once a remediation decision has been made, EPA prefers cleanups that achieve a risk closer to the more protective end of the range (for instance, 10^{-6}).

Baseline Assessments for Evaluating Risks to Human Health

The assessment of risks at Superfund sites across the country is a decentralized process. The risk assessment at each site is typically performed by contractors whose work is reviewed by EPA personnel. Recent risk assessments across sites can be combined for analysis, however, because they are generally conducted according to methodologies outlined in EPA's 1989 *Risk Assessment Guidance for Superfund, Volume I:*

Human Health Evaluation Manual (Part A), or RAGS. According to the RAGS, the baseline risk assessment conducted at each site has four objectives (U.S. EPA 1989a, 1–1):
- analyze the risks that might exist if no remedial actions or institutional controls were adopted at a site (that is, the "baseline risks") and help determine if actions are required at a site;
- provide information to help determine the maximum levels of chemicals that may remain on site, in order to protect public health;
- compare potential health impacts of remedial actions; and
- evaluate and document public health threats at sites.

The baseline risk assessment is part of the site characterization process in the remedial investigation/feasibility study (RI/FS). The RI/FS provides regional EPA decisionmakers with a quantitative assessment of human health risks at a site, a description of remedial action objectives, and an analysis of the alternatives proposed to reach these objectives. The reasoning behind the eventual course of action selected by EPA is contained in the ROD for each site.

The baseline risk assessment begins with the collection of site data, including samples taken to determine chemical concentrations at the site and to identify the potential chemicals of concern. Next an *exposure assessment* is conducted, where the risk assessor analyzes the contaminant data from the site, identifies exposed populations, determines potential exposure pathways, and estimates exposure concentrations and intakes by pathway. During the *toxicity assessment*, the next step, the analyst collects qualitative and quantitative information on the toxicity of the chemicals at the site, often using information from EPA's Integrated Risk Information System (IRIS). Finally, the information on exposure and toxicity is combined in *risk characterization models* to estimate cancer risks and noncancer hazard quotients for the chemicals and exposure pathways at the sites. The analysis in this chapter takes these risk assessments at face value and does not explore alternative risk assessment assumptions.

The *cancer risk* from a chemical is expressed as the incremental individual lifetime cancer risk from exposure to a substance from the site. *Chemical* cancer risks within a given exposure pathway are often aggregated to yield a *pathway* cancer risk, which would represent the incremental individual lifetime cancer risk for exposure to a set of chemicals via a given exposure scenario.

The *noncancer risk* of a chemical is conveyed by the *noncancer hazard quotient*, which equals the exposure level or intake of the chemical divided by its reference dose. The *reference dose* is an estimate of the exposure level that is likely to be without appreciable risks of harmful effects over a lifetime and is based on studies identifying the highest "no-observed-adverse-effect-level" (NOAEL) or the "lowest-observed-adverse-effect-level" (LOAEL).

The hazard quotients for multiple chemicals within an exposure pathway are then summed in the risk assessments to yield a *pathway hazard index*, which is meant to be a measure of noncancer risks associated with the pathway.

The data from the baseline risk assessments are designed in part to be used during the RI/FS process to establish remedial action objectives at the site, which are based both on the "applicable, or relevant and appropriate requirements" (ARARs) and the data from the risk assessments. Remedial actions must meet the ARARs of the Resource Conservation and Recovery Act (RCRA), Clean Water Act (CWA), Safe Drinking Water Act (SDWA), Clean Air Act (CAA), other federal statutes, and state environmental laws. ARARs generally fall into three categories (U.S. EPA 1988, 1–1):

- ambient or chemical-specific requirements that are generally health-based or risk-based numbers that translate into the amount or concentration of a chemical that may remain on site;
- performance or design requirements that limit the technologies or actions involving hazardous wastes at the site; or
- location-specific requirements that place restrictions on the concentration of hazardous substances at a site because of its location.

The remedial action objectives established are generally in terms of concentrations of particular chemicals that may remain after the remedial action is conducted.

Risk Assessment Pathways

Part of the decisions involved in conducting the baseline risk assessments lies in determining what pathways to evaluate to derive quantitative estimates of cancer and noncancer risks. Our database defines the pathways in risk assessments by a number of different category variables: time scenario of exposure, exposed population, age group, location of population, location of medium, exposure medium, and exposure route.

The *time scenario* variable refers generally to whether land use envisioned in the risk assessment corresponds to the current use or is related to a projected use in the future. Current land use is determined by the risk assessor according to site inspection data, zoning information, census data, and aerial photographs. Our designation of a pathway as a current or future scenario is determined by whether the risk assessment defined the pathway as current or future. Note that not all current risks are risk pathways that actually represent a risk today. Some assessments are based on current potential scenarios where the land use in an area does not change but other things may change, such as the size of a groundwater contamination plume so that wells not currently contaminated are assumed to become contaminated. These "current potential" risks are defined as current risks in our analysis if the risk assessors described them as such risks.

Future risks are generally those associated with changes in land use or activities. The guidance provided by the RAGS encourages risk assessors to consider a scenario where land that is currently not residential is brought into residential use in the future. The guidance document states:

> Because residential land use is most often associated with the greatest exposures, it is generally the most conservative choice to make when deciding what type of alternative land use may occur in the future. Assume future residential land use if it seems possible based on the evaluation of the available information. (U.S. EPA 1989a, 6–7)

Thus, future residential risks may be estimated at sites that are currently undeveloped or industrial and that have a low probability of future residential use. In our database of seventy-eight sites, there are thirty-five where future residential pathway risks occur despite the absence of any current residential risks exceeding the 10^{-6} cutoff level used in this analysis.

Exposed populations for which pathways are estimated include residents; workers; recreational users, such as swimmers or hunters; and trespassers. Though risk assessments are often conducted with very specific age group designations for the particular pathway described, we have (for this analysis) generally collapsed the different age groupings into adult (ages 18 and higher) and child (ages less than eighteen).

The risk assessment category for the *location of population* generally refers to where the particular population is exposed to the contaminant (for residents, location of population refers to where they live).

Location of medium refers to whether the contaminant for which the pathway is estimated is on-site or off-site.

Exposure medium describes in what medium the individual is exposed to the contaminant (such as air, groundwater, soil, or biota, that is, plants or animals containing chemicals that are later consumed by humans).

Exposure route details how a person comes into contact with the chemical. For example, soil contaminants may enter the body through ingestion or through dermal contact or through inhalation.

Categories of Cancer and Noncancer Pathways

In our analysis, we break down the description of cancer and noncancer pathways by risk assessment categories. Determining the relative magnitudes of current versus future risks is important in distinguishing how estimates of human health risks at Superfund sites are affected by assumptions about future land use. Designating whether risks involve residents, workers, recreational users, or trespassers is a necessary step in analyzing the efficacy of different policy options for reducing human health risks. Similarly, analyzing whether the populations exposed are on-site or off-site and whether the contaminants are on-site or off-site is a necessary part of evaluating the impact of remedies at Superfund sites.

We also analyze the contribution of specific chemicals to the risks posed. Since uncertainty may exist over the toxicity of particular chemicals, consideration of the relative frequency of these chemicals at sites and their estimated contribution to pathway risks may help determine where additional resources could be devoted to defining the risks of these chemicals or developing remedies to deal with particular types of contaminants.

After a brief description below of the database construction, we turn in the subsequent section to an analysis of the cancer and noncancer risks by their category of risk assessment groupings.

DATA CONSTRUCTION

The data necessary to analyze fully the human health risks and estimated remediation costs associated with Superfund sites are spread across the country in the site administrative records maintained at the ten EPA regional offices. Though EPA has a central repository for RODs in Washington, the background documents that lead up to each ROD are only available at the regional level. We sent researchers to these regional EPA offices with instructions to collect for each site in our sample the complete baseline risk assessment, extended excerpts from the RI/FS, the complete ROD, and any modifications to the ROD.

While RODs contain extensive details of the pathway risks estimated in the baseline risk assessment, we considered it essential to go beyond the ROD to collect additional data for several reasons. The ROD does not include the full baseline risk assessment, which provides information on the parameter values used in the risk assessment calculations, the reasonable maximum exposure (RME) point concentrations of the particular chemicals employed to calculate pathway risks, and in many cases the average concentrations of these chemicals. ROD risk summaries are also sometimes presented so that pathway risks are combined to form the risks to a particular population, or so that risks are presented for different chemicals but not for particular pathways. These risks would be difficult to analyze at the pathway level if one did not go back to the baseline risk assessment. In addition, the details necessary to link pathway risks to particular populations in order to develop population risks at Superfund sites are often only found in the baseline risk assessment or RI/FS documents.

Our research assistants gathered information from the regional offices on all sites that had a ROD signed during 1991 or 1992. These two years were chosen since the risk assessments conducted at these sites were likely to have been performed using the methodology outlined in EPA's 1989 *Risk Assessment Guidance for Superfund* (U.S. EPA 1989a, 1989b). There were a total of 276 RODS signed during this period at a total of 266 sites. We have entered information on human health risks into our database for seventy-eight sites for analysis in this chapter. The distribution of these sites across the country is similar to the distribution of total NPL sites across the country.[2] Further work on the population risks and cost-risk trade-offs made at Superfund sites will focus on an expanded number of sites with RODs signed in 1991 or 1992.

We developed risk coding sheets that allowed research assistants to enter into the database human health risk assessment information from the baseline risk assessment, the RI/FS, and the ROD. We collected information such as chemical concentrations, risk assessment parameters used in the models to derive cancer and noncancer risks at the chemical and pathway levels, and descriptions of the different pathways (for instance, what scenario, exposed population, and exposure medium were associated with a particular pathway risk?). We checked the data entered in two ways. First, we compared the database figures against the original documents. Second, we did an independent calculation of the pathway risks through the use of the chemical concentration information and risk assessment parameters collected, which we compared to the figures in the original documents.

The baseline risk assessment and ROD for a single site may contain pathway risks for pathways associated with extremely small risks. Our decision rules for entering risks into the database were as follows. The RODs served as the first source for pathway risk data. If a ROD contained risk information on all cancer pathway risks that were at least 1×10^{-6} and noncancer pathway risks with a hazard index greater than or equal to one, then we entered all the ROD risks that met the following risk cutoff levels: 1×10^{-7} for cancer pathway risks; 1×10^{-8} for cancer risks arising from an individual chemical within a pathway; 0.1 for the noncancer pathway hazard index; and 0.01 for the noncancer hazard quotient for each chemical. If the ROD did not present the minimum pathway risk data we required, we turned to the baseline risk assessment and entered the risk data according to the above decision rules. If the ROD risks were presented in forms other than by risk pathways (for instance, if only risks by chemical were presented), then the baseline risk assessment figures were employed. For each site, data on chemical concentrations and risk assessment parameters came from the baseline risk assessment.

Pathway risks smaller than the 1×10^{-6} figure, which is often cited as a cutoff for EPA action, were collected because the aggregation of these smaller risks may aid in the calculation of population risks when the human health assessment figures are combined with census population figures surrounding sites. For our initial analysis in this chapter of the risk data used by regional EPA officials to make their remediation decisions, however, we have analyzed for the risk information in our database all cancer pathway risks greater than or equal to 1×10^{-6}, all noncancer pathway risks with hazard indices greater than or equal to one, and all chemicals associated with these pathways. These cutoffs eliminated one site in our sample at which the no action alternative was chosen where pathway risks were less than these thresholds, so we analyze in the following section cancer and noncancer risks at seventy-seven Superfund sites.

RISK PATHWAY MECHANISMS

Analyzing the exposure pathways at Superfund sites sheds light on the manner in which the risks arise, which then informs the debate over how remedies should be selected. Do the risks arise from groundwater contamination, soil ingestion, or other mechanisms? Do current uses of the land comprising the site give rise to the risk, or is it some future use that has not yet occurred? Questions such as these are of obvious interest from the standpoint of risk analysis, since the nature of the pathway generating the risk will influence the degree of exposure and the duration of exposure. Examination of the risk pathways, however, is also instructive from a policy standpoint.

EPA has a variety of policy options that it can adopt with respect to Superfund cleanup. If the main risk is that from groundwater contamination, households could switch to alternative water supplies to avoid the risk associated with this particular pathway. In addition to various stringent options involving treatment of the waste, there are also intermediate options that include the use of restrictions on the use of the land, capping and fencing the site, and similar measures that may not eliminate the presence of the chemical, but that would eliminate certain risk pathways as being influential.

An examination of the distribution of pathways is instructive to get a sense of the frequency with which alternative risk exposure mechanisms are operative. However, one should be cautious in proceeding from a pathway count to making inferences about the total level of the risk associated with a particular grouping of pathways. The risk associated with a set of pathways is governed not only by the number of such pathways but also by the magnitude of the risk associated with them. Pathways for which there is a high probability of an adverse outcome consequently pose greater risk than those with a lower probability. (Refinements of the analysis to include recognition of the magnitude of risk associated with pathways, as opposed to simply the number of pathways, appear in this chapter's section on these associated risk levels.)

Another determinant of the risk is not only the probability but also the size of the exposed population. Population-weighted risk levels do not comprise a component of the EPA analysis and thus do not explicitly enter the analytical basis for Superfund policymaking. In subsequent work, we plan to explore how recognition of the population levels exposed to different categories of risk would influence the assessed risk levels. What is available as part of the EPA analyses is information on the type of population group exposed, such as residents or recreational users, and we analyze the different risk levels to these population groups below.

Distribution of Pathways by Risk Assessment Categories

Table 1 provides a comprehensive overview of the distribution of the risk pathways by various categories of analysis. The columns of statistics in the table provide the pertinent breakdowns within the risk assessment categories for all 1,430 pathways, for the 1,015 cancer pathways, and for the 415 noncancer pathways in the sample.

The first distinction in the table, which is perhaps the most salient result of the study, pertains to the breakdown between risks arising from current uses of the land and risks arising from future uses. This distinction pertains not to the time period of the risk but rather to the nature of the context in which the risks will arise. For example, future risks to current residents are generally captured under the "current"timeframe designation, but new uses, such as the decision to build a residential area on land that is now a Superfund site, would be a "future" risk. The striking result of Table 1 is that the great majority of the risk pathways pertain to such future risk exposures as opposed to risks associated with current uses. Overall, 70% of the cancer pathways, 79% of the noncancer pathways, and 72% of the total pathways pertain to future as opposed to current uses.

Table 1. Distribution of Pathways by Risk Assessment Categories (percent)

Risk assessment category	Total pathways (N = 1,430)	Cancer pathways (N = 1,015)	Noncancer pathways (N = 415)
Scenario			
current	27.8	30.5	21.0
future	72.2	69.5	79.0
Exposed population type			
residential	73.2	71.2	78.1
worker	17.4	17.8	16.4
recreational	3.6	3.8	3.1
trespasser	5.8	7.2	2.4
Age group			
adult	62.7	65.3	56.4
child	37.3	34.7	43.6
Location of population			
on-site	69.2	70.2	66.7
off-site	23.2	23.3	22.9
not indicated	7.6	6.5	10.4
Location of medium			
on-site	79.6	80.3	77.8
off-site	13.7	14.3	12.3
not indicated	6.7	5.4	9.9
Exposure medium			
air (from soil)	9.0	9.0	8.9
air (from water)	9.0	10.4	5.3
soil	33.6	38.2	22.2
groundwater	37.2	30.8	52.8
surface water	1.0	1.1	0.7
sediment	5.2	5.7	3.9
biota	3.6	2.9	5.3
structures	0.1	0.2	–
sludge	0.8	0.9	0.5
combination	0.3	0.4	–
leachate	0.1	0.2	–
mothers' milk	0.3	0.2	0.5
Exposure route			
ingestion	58.4	53.7	69.9
dermal contact	22.6	25.7	14.9
inhalation (vapor phase chemicals)	13.0	14.6	9.4
inhalation (dust)	5.7	5.8	5.3
dermal contact and inhalation	0.3	0.2	0.5

Of the exposed population types, the most important in terms of the risk pathways is that of residential populations. Approximately three-fourths of all pathways pertain to residential populations, with the next most important group being workers, for whom only 17% of the pathways are pertinent. Recreational users, such as those who fish in streams on Superfund sites, account for a very small fraction of all the risk pathways.

In terms of the age distribution of those affected by the risk pathways, most of the risk pathways (over 60%) are to adult populations, and just over one-third of the risk pathways pertain to children (that is, those under 18 years of age). The main difference between these figures and the overall age distribution of the U.S. population is that whereas 37% of the risks are to the child age population, this group comprises only 26% of the U.S. population overall.[3] Thus, the pathways affecting children occur almost 1.5 times as often as the representation of children in the population.

The location where the risks arise is also of substantial interest, particularly as it relates to the potential efficacy of policy options that limit future uses of land at or near Superfund sites. Both the location of the populations and the location of the medium (that is, the location of the medium from which the risk arises) are heavily concentrated toward on-site risks. Of the total pathways, 69% pertain to risks to on-site populations; 80% of the media associated with the pathways pertain to on-site media. The particular media that appear to be most prominent are soil and groundwater, as each of these accounts for over one-third of all pathways. The other relatively important exposure media are air (from soil), air (from water), and sediment, each of which accounts for 5% to 10% of all pathways. If the two air pathway mechanisms are aggregated, they account for almost one-fifth of all pathways.

The final component of Table 1 lists the exposure route by which the risk arises. The dominant exposure route is that of ingestion, such as drinking contaminated groundwater or ingesting dirt, where this category gives rise to 58% of all pathways. Dermal contact accounts for 23% of all of the different exposure routes, and inhalation of vapor phase chemicals and dust are next in importance.

For the exposure routes as well as for most of the other components of the table, the distribution of pathways is fairly similar for both cancer pathways and noncancer pathways. The major distinctions are that the noncancer pathways play a more prominent role in the future risk scenarios, are more likely to affect residential populations, are less likely to affect adults, are more likely to involve groundwater exposure rather than soil, and are more likely to arise from ingestion rather than dermal contact.

Distribution of Pathways by Scenario and Exposed Population

Table 2 analyzes the distribution of the exposed population and the location of the exposed population for each of the two land use scenarios. Overall, the great majority of the pathways are accounted for by residents based on future risk scenarios, which account for 59% of the pathways. In contrast, current risks to current residents, as well as future populations in current residential areas, account for only 14% of the risk. The next most prevalent category, that of workers, also has a greater number of pathways for future scenarios as opposed to current time frames, but the difference is not as stark as in the case of the residential risks.

There is also a substantial difference in the character of the risks with respect to their population location and the time frame for the analysis. On-site risks under current scenarios account for 15% of the pathways, which is only somewhat greater than the current off-site risks of 11%. For the future-based scenarios, however, on-site risks escalate to account for 54% of the pathways, which is more than four times as great as the 12% of the pathways due to future off-site risks. Overall, 90% of current pathways are on-site residential pathways, 59% of future pathways are on-site residential, and future on-site residential pathways account for 43% of all pathways in the sample. *The dominant exposure to risks consequently arises from the expected future residential exposures on Superfund sites.*

In the case of residents, the chief risks arise from ingestion of either groundwater or soil. Resident ingestion of groundwater accounts for a quarter of all total pathways. Although Superfund anecdotes frequently highlight the importance of children who eat dirt, it is noteworthy that ingestion of soil plays a much greater proportional role in the risk pathways for workers than it does for residents. Dermal contact with soil and ingestion of groundwater also account for a substantial share of the risks to workers and a significant share of all total pathways.

Table 2. Distribution of Pathways

Distribution by exposed population type				
Scenario	Residential	Worker	Recreational	Trespasser
current	13.85	6.43	1.96	5.52
future	59.37	10.91	1.68	0.28
Distribution by population location				
Scenario	On-site	Off-site		Not indicated
current	15.38	10.91		1.47
future	53.78	12.31		6.15

Note: All figures are a percentage of total pathways, where $N = 1{,}430$.

The risk pathways for recreational users and trespassers account for a very small percentage of all pathways in the sample, but the distribution within these groups is nevertheless of interest. The primary risk to recreational users is that from dermal contact with soil, with ingestion of soil and inhalation of the vapor phase chemicals from soil being next in importance. For trespassers, the major risks are from ingestion of soil and ingestion of sediment, which together account for almost half of all the risks to trespassers.

RISK LEVELS ASSOCIATED WITH PATHWAYS

Although examination of the distribution of the number of pathways is instructive in providing an assessment of the mechanisms by which risks arise, the level of risks associated with the pathways is also of substantial consequence. Some pathways involve intense exposure to very hazardous chemicals, whereas others involve more minimal exposure levels. In this section we explore different features of the aspects of the risk distribution of the pathways included in our sample.

Distribution of Risk Pathways by Risk Levels

A useful starting point is the distribution of the risk pathways by risk level, which appears in Table 3. The top panel presents the distribution of the cancer risk pathways for different risk ranges. Over half of the cancer pathways in the sample pertain to risk levels below 10^{-4}. However, it is quite striking that many of the pathways involve considerable risks, with eighteen of the pathways posing cancer risks in excess of 1/10. Health risk levels of this kind are not unprecedented. For example, cigarettes pose a lifetime cancer risk of about 1/3.[4] Some of the large pathway risks are very large risks by comparison with the targets of most government risk policies. The risk threshold for most federal risk policies is either one in 100,000 or one in 1,000,000, and even job fatality risks for blue collar workers are only on the order of one in 10,000. In contrast, many of the risk pathways in Table 3 are associated with risk levels orders of magnitude greater than these levels.

Such large risks arise in part because of particular risks associated with some extremely hazardous sites. The Westinghouse site in California is perhaps most noteworthy in that it accounts for four of the top ten cancer risk pathways. The risks arising at that site are from high concentrations of PCBs. There are multiple pathways because there are different population groups exposed to these risks. As in the case of the overall distribution of the risk pathways, it is the future scenarios that

Table 3. Distribution of Risk Pathways by Risk Level

Number of cancer pathways by risk range

1E–6 to 1E–5	1E–5 to 1E–4	1E–4 to 1E–3	1E–3 to 1E–2	1E–2 to 1E–1	1E–1 to 1
301	348	205	103	40	18

Number of noncancer pathways by hazard index

1 to 10	10 to 100	100 to 1000	> 1000
254	126	22	13

Note: The notation 1E–N indicates 1×10^{-N}.

play a dominant role. The risk pathways responsible for the high risk ranking of the Westinghouse site are the risks posed to adults on-site (dermal exposure to soil), workers on-site (dermal exposure to soil), children on-site (soil ingestion), and on-site resident children (dermal exposure to soil), where all these risks pertain to future risk scenarios as opposed to current risk pathways. This prominence of future risk pathways extends beyond this particular site. All of the top thirteen pathway risks are associated with future as opposed to current risk scenarios.

The bottom panel of Table 3 also presents the distribution of the relative risk levels for the noncancer pathways. These figures pertain to the hazard index associated with noncancer risk. Chemicals differ in the potency of their health effects, so that one should be very careful in interpreting any aggregation of these statistics across chemicals. For the most part, the chemical exposures are less than ten times greater than the hazard index threshold, but in some cases there are extreme chemical exposures that are 1,000 times as great as the reference dose for the chemical.

Magnitude of Pathway Cancer Risks

Table 4 provides a comprehensive overview of the magnitude of the risks for all the principal risk assessment categories. Whereas Table 1 provided information on the percentage of pathways in each of the various risk assessment categories, Table 4 provides information on the risk levels associated with these categories, where the main statistics of interest are mean risk levels and the median risk. Because of the influence of the very high risk outliers, the mean risks are consistently larger than the median risk levels, but the overall relationships are roughly similar across risk assessment categories.

Earlier, we found that future risks dominated in terms of the frequency with which these pathways occurred. This greater occurrence rate is reinforced by the higher risk levels associated with future risk pathways, as the risk levels per future risk pathway exceed the risk levels per current use pathway by a factor of five for the means and a factor

Table 4. Pathway Mean and Median Cancer Risks by Risk Assessment Category

Risk assessment category	Mean	Standard deviation	Median	N
Overall	7.5E–3	5.6E–2	3.1E–5	1,015
Scenario				
current	2.2E–3	1.2E–2	1.9E–5	310
future	9.9E–3	6.7E–2	4.0E–5	705
Exposed population type				
residential	9.2E–3	6.5E–2	3.7E–5	723
worker	4.7E–3	2.9E–2	2.6E–5	180
recreational	2.8E–3	7.6E–3	4.0E–5	39
trespasser	2.8E–4	9.1E–4	9.0E–6	73
Age group				
adult	8.6E–3	6.3E–2	3.1E–5	663
child	5.5E–3	4.1E–2	3.2E–5	352
Location of population				
on-site	8.3E–3	5.9E–2	3.4E–5	712
off-site	6.4E–3	5.4E–2	3.0E–5	237
not indicated	3.1E–3	1.6E–2	2.1E–5	66
Location of medium				
on-site	8.8E–3	6.2E–2	3.5E–5	815
off-site	2.1E–3	1.4E–2	2.0E–5	145
not indicated	3.7E–3	1.7E–2	1.9E–5	55
Exposure medium				
air (from soil)	3.8E–3	1.9E–2	1.7E–5	91
air (from water)	1.5E–3	9.2E–3	2.9E–5	106
soil	6.5E–3	6.1E–2	2.1E–5	388
groundwater	1.2E–2	6.8E–2	1.6E–4	313
surface water	2.7E–4	7.6E–4	4.4E–5	11
sediment	7.1E–4	2.3E–3	1.9E–5	58
biota	2.7E–2	8.5E–2	4.4E–5	29
structures	1.0E–4	1.4E–4	1.0E–4	2
sludge	3.1E–3	8.3E–3	2.0E–4	9
combination	9.0E–6	3.0E–6	9.0E–6	4
leachate	5.0E–6	0.0	5.0E–6	2
mothers' milk	7.0E–2	9.9E–2	7.0E–2	2
Exposure route				
ingestion	9.2E–3	5.6E–2	4.5E–5	545
dermal contact	8.2E–3	7.3E–2	2.1E–5	261
inhalation (vapor phase chemicals)	3.2E–3	1.7E–2	3.1E–5	148
inhalation (dust)	4.5E–4	1.5E–3	1.7E–5	59
inhalation and dermal	7.1E–5	3.9E–5	7.1E–5	2

Note: The notation 1E–N indicates 1×10^{-N}.

of two for the medians. Not only are the future scenario risk pathways in Superfund human health risk assessments much more prevalent than existing pathways, but *when future pathways do occur in the analysis, they have a much higher risk*. Both the frequency and the severity of the risks of the future risk pathways are consequently greater.

The chief risks facing exposed populations are those incurred by residential populations and workers. The trespasser risk probability levels are quite small. Coupling these low risk levels with the low frequency of trespasser pathways (shown in Table 1) suggests that there may be little danger to trespassers from policy options that would not treat or remove the chemicals but simply restrict the use of the site in the future. Even without fencing or other barriers, the overall frequency of the trespasser pathways and the severity of the risks associated with these pathways is not as great as for other populations at risk.

The magnitude of the risks associated with different age groups is also somewhat contrary to general beliefs that primarily children face the greatest risks. The *level* of the risks faced by adults is greater than that facing children, although the median risks are virtually identical.

In terms of the location of the risks, the on-site populations face the greatest risks and the on-site media pose the greatest risks. Preventing future development of the site or use of the site for other purposes would, for example, eliminate the most severe risks that arise. Of the various exposure media, many posing the largest risk (such as mothers' milk and biota) are associated with very few pathways. The most prevalent pathways, those linked to soil and groundwater, pose mean cancer risks on the order of 0.01 and 0.001. These estimated risks are several orders of magnitude larger than those driving many other federal risk regulation efforts.

Dimensions of the Distribution of Cancer Pathway Risks

Table 5 distinguishes the different risk levels according to the principal dimension that influences the pathway distribution: the time frame for the risk scenario. It indicates for each of the time frames what the risk levels are for different exposed populations and different population groups. The largest population group by far is the on-site residential population for future risk scenarios, and this population group accounts for 431 pathways and also faces the greatest risk, which is 1.1×10^{-2}. This risk level is several times greater than the risks facing current on-site residents. The EPA risk analysis consequently assumes not only that on-site resident pathways will be much more prevalent than they are now but that these pathways also will pose greater risks thanthose faced by current on-site residents. Future workers and future

Table 5. Distribution of Cancer Pathway Risks by Scenario, Population Location, and Exposed Population

		Exposed population type			
Population location		Residential	Worker	Recreational	Trespasser
Current scenario					
on-site	count	52	66	13	42
	mean	1.1E–3	2.8E–3	1.8E–3	2.9E–4
	median	3.7E–5	2.1E–5	4.3E–5	1.2E–5
off-site	count	82	7	9	23
	mean	3.6E–3	4.4E–4	1.1E–4	3.6E–4
	median	1.7E–5	6.0E–5	1.4E–5	2.6E–5
not indicated	count	11	1	–	4
	mean	8.3E–3	1.0E–6	–	9.0E–6
	median	7.9E–6	1.4E–6	–	5.6E–6
Future scenario					
on-site	count	431	91	15	2
	mean	1.1E–2	7.1E–3	5.6E–3	5.0E–6
	median	4.0E–5	2.9E–5	8.8E–5	4.8E–6
off-site	count	106	8	2	–
	mean	1.1E–2	2.2E–4	5.5E–4	–
	median	6.0E–5	3.1E–5	5.5E–4	–
not indicated	count	41	7	–	2
	mean	2.5E–3	1.4E–3	–	3.0E–6
	median	9.3E–5	2.0E–5	–	3.1E–6

Note: The notation 1E–N indicates 1×10^{-N}.

recreation users also will face greater risks than their current scenario counterparts, although future trespassers will face a lower risk. Overall, trespassers face the lowest risk levels in the sample, and future trespassers are not only infrequent but face an extremely small risk level.

The population locations of the different parties at risk are reflective of the nature of the exposed populations. Since future residents are in many cases on-site, the on-site future scenario group accounts for the largest number of pathways as well as the highest average risk level. The future off-site risk probabilities are comparable to the future on-site risks. It is also interesting that the on-site and off-site risks for the current scenarios are similar as well. The main difference is not between the on-site and off-site location but rather whether it is a current or future scenario, as the future on-site and off-site risks are at least four times larger than the risk levels assumed based on current scenarios. Thus, the EPA analyses are predicated not only on an assumption that future scenarios involving exposed populations will be the dominant pathways, but also that these future scenarios will expose the population in a manner that will give rise to much larger risks than those now faced based on the current structure of economic development in the area.

Risk-Weighted Shares of Cancer Pathway Risks

Analysis of the frequency of risk pathways gives a sense of how often the pathways are pertinent; consideration of the risk levels associated with the pathways indicates the magnitudes of the risk per pathway. However, the overall level of the risk that will be generated at a Superfund site will reflect the combined influence of the frequency with which particular types of risk pathways occur as well as the levels of the risk associated with different types of pathways. A fuller analysis would also recognize the size of the populations exposed to the risk. Note, however, that the current EPA mandate dictates that the risk posed from hazardous waste sites should be presented as risks to the individual, not to the population. As stated in the National Oil and Hazardous Substances Pollution Contingency Plan of 1990,[5] "acceptable exposure levels for carcinogens are generally concentration levels that represent an excess upperbound lifetime cancer risk to an individual of between 10^{-4} and 10^{-6}." As a result, baseline risk assessments focus on individual risk.

To convey information concerning both the frequency and magnitude of pathway cancer risks, Table 6 provides statistics on the risk-weighted shares of the different cancer risk pathways. Rather than simply determining the fraction of the pathways represented by particular types of exposures, such as risk to future generations, each of these pathways is weighted by the total magnitude of the risk estimated for that pathway and the risk-weighted pathways are then summed for the entire sample. The statistics in Table 6 provide information on the percentage of the total risk-weighted pathways accounted for by each pathway type.

The principal purpose of combining the influence of the frequency of pathway occurrence with the magnitude of the risk is to generate a hybrid of the two influences discussed above. For example, in the case of future risk scenarios, we found that future risk pathways were not only more prevalent than pathways based on current risk scenarios, but that these pathways posed a greater risk level per pathway as well. The compounding of these influences is borne out in the statistics in Table 6, as 91% of all total cancer pathway risks are attributable to future risk scenarios. This emphasis on future risks is much greater than the unweighted share of future pathways, which we found in Table 1 to be only 72%.

The other statistics in the table are presented for total cancer pathway risks, future cancer pathway risks, and current cancer pathway risks. In terms of the distribution of risks, by far the largest risk share is for adults for current pathways and for future pathways.

A very strong contrast arises with respect to risks to the various exposed populations. Residential populations account for 66% of the

Table 6. Risk-Weighted Shares of Cancer Pathway Risks (percent)

Risk assessment category	Total cancer pathway risk	Future cancer pathway risk	Current cancer pathway risk
Scenario			
current	8.8	–	100.0
future	91.2	100.0	–
Age group			
adult	74.9	74.5	78.5
child	25.1	25.5	21.5
Exposed population type			
residential	87.3	89.4	65.6
worker	11.0	9.4	27.7
recreational	1.4	1.2	3.6
trespasser	0.2	0.0	3.1
Location of population			
on-site	77.5	81.0	40.9
off-site	19.9	17.4	45.4
not indicated	2.7	1.6	13.7
Exposure medium			
air (from soil)	4.5	4.2	7.3
air (from water)	2.1	2.0	3.1
soil	32.9	34.2	19.4
groundwater	47.7	49.1	32.8
surface water	0.0	0.0	0.0
sediment	0.5	0.5	1.0
biota	10.0	7.5	36.5
structures	0.0	0.0	0.0
sludge	0.3	0.3	0.0
combination	0.0	0.0	0.0
leachate	0.0	0.0	0.0
mothers' milk	1.8	2.0	0.0
Exposure route			
ingestion	65.4	64.6	74.1
dermal exposure	28.0	29.2	15.6
inhalation (vapor phase chemicals)	6.2	5.9	9.4
inhalation (dust)	0.3	0.3	0.9
ingestion and dermal	–	–	–
inhalation and dermal	0.0	–	–

current cancer pathway risks, and this figure escalates to 89% for future risk pathways. Similarly, exposures to workers account for 28% of current cancer pathway risks, and this figure drops to only 9% for future risk pathways.

The location of the populations affected by these risks also changes dramatically depending on the time frame for the risk scenario. The percentage role of on-site population risks rises from 41% to 81% when one moves to the future risk scenarios, and the role of off-site risks drops from 45% to 17%. The implications of the exposed population and location of population results is that future risk scenarios put a much greater weight on risks posed to on-site residential areas. Of the exposure media listed in Table 6, the most noteworthy pattern is that groundwater risks account for almost half of the future cancer risk pathways, as contrasted with about one-third of these pathways for current cancer risk assessments. The role of biota drops substantially for future risk pathways.

Maximum Risk Pathways

Although the overall risk-weighted pathways provide the most comprehensive assessment of the risk level from the standpoint of policy analysis, for policy decisions it may be the maximum risk at particular sites that drives the policy choice. These risks are likely to be most salient in the policy debate and are frequently the object of considerable emphasis in the discussion in Superfund documents. Table 7 summarizes the distribution of the maximum risk pathways for different sites, where these breakdowns are provided according to the same categories as were used to characterize the overall percentage distribution of pathways in Table 1 and the percentage distribution of the risk-weighted pathways in Table 6.

In the case of the current versus future risks, the results for the maximum site pathways are intermediate between the overall pathway distribution and the risk-weighted pathways. Future risk pathways accounted for 72% of all pathways and 91% of all risk-weighted pathways. Not surprisingly, the 79% share for the maximum site pathways in future risk scenarios lies between these two estimates because sites associated with the maximum risk will receive greater weighting when computing the risk-weighted pathway share.

The policy implications of examining only the maximum risk pathways do not always parallel those that would derive from an examination of all cancer pathways. In terms of total pathways, soil-related risks account for 34% of the pathways, and groundwater accounts for 37% of the pathways. The risk-weighted pathway shares are only slightly different, as soil has a 33% risk-weighted share and groundwater has a 48% risk-weighted share. If, however, one examines the maximum site pathway, the role of the soil pathway drops to 20%, and the groundwater share rises to 65%, far in excess of the overall risk-weighted share of cancer risk pathways. The public debate may have placed inordinate attention on the role of groundwater hazards since these risks are frequently the maximum site pathways at sites. A more comprehensive analysis that takes into account the frequency of pathways as well as their severity suggests that the role of groundwater contamination is

Table 7. Distribution of Maximum Site Pathways by Risk Assessment Categories (percent)

Risk assessment category	Cancer pathways ($N = 86$)	Noncancer pathways ($N = 73$)
Scenario		
current	20.9	13.7
future	79.1	86.3
Exposed population type		
residential	84.9	95.9
worker	15.1	2.7
recreational	–	–
trespasser	–	1.4
Age group		
adult	86.0	41.1
child	14.0	58.9
Location of population		
on-site	69.8	64.4
off-site	23.3	28.8
not indicated	7.0	6.8
Location of medium		
on-site	83.7	79.5
off-site	9.3	13.7
not indicated	7.0	6.8
Exposure medium		
air (from soil)	3.5	6.8
air (from water)	7.0	1.4
soil	19.8	6.8
groundwater	65.1	76.7
surface water	–	1.4
sediment	2.3	1.4
biota	2.3	4.1
structures	–	–
sludge	–	–
combination	–	–
leachate	–	–
mothers' milk	–	1.4
Exposure route		
ingestion	77.9	89.0
dermal contact	11.6	2.7
inhalation (vapor phase chemicals)	8.1	4.1
inhalation (dust)	2.3	4.1
dermal contact and ingestion	–	–
dermal contact and inhalation	–	–

Note: The count for any given category may not equal the number of sites for two reasons. First, some sites do not have a unique maximum risk. In this case, all maximum risks were included. Second, some sites may lack either carcinogenic risk or noncarcinogenic risk.

much less than would be suggested by the maximum risk pathways alone. Note, however, that groundwater exposure pathways have the potential to affect larger populations than soil exposure pathways.

The breakdowns in Table 8 carry further the distinctions for maximum pathway cancer risks by examining the risks for different population groups. The top panel, which pertains to the current risk scenarios, indicates that there will be eighteen current maximum risk pathways. Mean risks for these pathways are high—varying from 0.003 to 0.09—although the number of sites represented is often quite small. The most frequent pathways are for on-site locations. There are sixty-seven future maximum risk pathways in the bottom panel of Table 8, where the most important in terms of frequency is that pertaining to on-site residents. The mean risks associated with each of these maximum risk pathways are virtually indistinguishable for on-site residents, on-site workers, and off-site residents.

CONCLUSION

Most of the political pressures that generated the impetus for the Superfund program arose because of the concern of existing populations for the risks that they believe these sites currently pose. Consideration of the risk assessments for Superfund sites indicates, however, that it is not the existing risks that are most salient. Rather, the dominant risks arise from future risk scenarios that generally involve alternative uses of the land. Indeed, these future risks account for 90% of all the risk-weighted pathways for the Superfund sites in our sample. Chief among these future risks is that of future residents living on-site. The underlying assumption driving the EPA risk analyses is that there will be new residential areas on existing future Superfund sites where there are currently no such residential areas.

Analysis of the structure of risks is of fundamental importance with respect to the choice of different possible modes of government intervention. If some mechanism were available that could eliminate these future risks, such as the use of various use restrictions and containment options, then the great preponderance of the risks analyzed in human health assessments at Superfund sites would be eliminated. Indeed, examination of the risk pathways suggested that many of the risks likely to remain with such containment and land-use restriction options, such as that to trespassers, are very low even without adopting policies, such as fencing, to reduce these risks.

Although many observers have attempted to dismiss Superfund risks as being trivial, many of the estimated hazards are quite substantial.

Table 8. Distribution of Maximum Pathway Cancer Risks across Sites by Scenario, Exposed Population, and Population Location

Current scenario		Population Location ($N = 18$)		
Exposed population type		On-site	Off-site	Not indicated
Residential	# of sites	6	3	1
	mean risk	7.1E–3	4.4E–2	9.1E–2
	median	1.3E–3	1.1E–3	9.1E–2
Worker	# of sites	7	1	–
	mean risk	1.2E–2	2.8E–3	–
	median	1.1E–2	2.8E–3	–
Recreational		–	–	–
Trespasser		–	–	–
Future scenario		Population Location ($N = 67$)		
Residential	# of sites	42	16	5
	mean risk	6.7E–2	5.7E–2	1.9E–2
	median	2.2E–3	1.1E–3	1.5E–3
Worker	# of sites	5	–	–
	mean risk	9.7E–3	–	–
	median	2.0E–3	–	–
Recreational		–	–	–
Trespasser		–	–	–

Note: The notation 1E–N indicates 1×10^{-N}.

Although the EPA risk threshold for considering a pathway risk is generally a lifetime cancer risk of one in a million, the mean risk level associated with pathways is typically several orders of magnitude larger than this threshold. Moreover, these mean risk levels pertain not only to the site generally, but also to a variety of different kinds of pathway mechanisms and different groups of exposed populations at the site. Taken at face value, these risk assessments suggest Superfund risks exceed estimated risks for other federal cancer regulation efforts. Thus, even if one chooses to disregard some pathway mechanisms as being unlikely, the overall scale of the risks is sufficiently large that such casual dismissals of Superfund risks based on anecdotal evidence are not warranted.

Ultimately, to form a reliable assessment of the merits of the Superfund program and possible alternative modes of government intervention, one needs to refine the risk analysis in a variety of ways. Our study considered the frequency of different types of risk and their associated risk levels, but it did not address the magnitude of the populations affected or the cost of achieving risk reductions. Moreover, one would want to assess the effect of different policy alternatives on cost and risk in the usual type of policy analysis. The current legislative focus of the Superfund program is risk-oriented, as is the case of most other governmental human health regulations. Given this emphasis, even a partial analysis that considers only the estimated risks at Superfund sites is especially instructive in illuminating a stated target of this program.

ACKNOWLEDGMENTS

We would like to thank Scott Rehmus for directing the research staff, Chris Dockins for computer assistance, Kristen Blann for editorial contributions, and Joey Ams, Jason Bell, Kristen Blann, Chris Chiang, Elizabeth Gregory, Jahn Hakes, Lisa Larson, Robert Malme, Jason Parsley, Zac Robinson, and A. Este Stifel for help in constructing the database. The Office of Policy Planning and Evaluation provided complete financial support under CR-817478-02. Dr. Alan Carlin, the contract officer, provided advice and guidance throughout this project. A fuller version of this chapter appeared in *Ecology Law Quarterly*, Vol. 21, No. 3, 1994.

ENDNOTES

[1] Ultimately, the bill was not reauthorized in 1994.

[2] The percentage of site distributions across EPA Regions 1 through 10 of the seventy-eight sites analyzed in this study are, respectively, 3.8%, 29.5%, 11.5%, 11.5%, 23.1%, 3.8%, 7.7%, 1.3%, 5.1%, 2.6%. The percentage of site distributions across Regions 1 through 10 of all RODs signed through 1992 are, respectively: 6.5%, 17.3%, 14.5%, 11.4%, 19.8%, 6.9%, 5.5%, 6.0%, 7.8%, 4.5%. The percentage of site distributions across Regions 1 through 10 of nonfederal sites on the NPL are, respectively: 6.8%, 17.3%, 13.6%, 13.2%, 23.0%, 6.0%, 5.0%, 3.1%, 7.4%, 4.4%.

[3] U.S. Department of Commerce 1992, 19, table 19. Statistics are for 1991.

[4] From Viscusi 1992.

[5] The National Oil and Hazardous Substances Pollution Contingency Plan of 1990 is codified in the *Code of Federal Regulations*: 40 CFR Section 300.430 (e)(2)(I)(A)(2) 1993. See also 55 *Federal Register* 8665–8865, March 8, 1990.

REFERENCES

Acton, Jan Paul, and Lloyd S. Dixon. 1992. *Superfund and Transaction Costs: The Experiences of Insurers and Very Large Industrial Firms.* RAND/R-4132-ICJ. Santa Monica, California: RAND.

Burmaster, David E., and Robert H. Harris. 1993. The Magnitude of Compounding Conservatisms in Superfund Risk Assessments. *Risk Analysis* 13 (2): 131–134.

Doty, Carolyn B., and Curtis C. Travis. 1989. The Superfund Remedial Action Decision Process: A Review of Fifty Records of Decision. *Journal of the Air Pollution Control Association* 39 (12): 1535–1542.

ENVIRON Corporation. 1993. *A Comparison of Monte Carlo Simulation-Based Exposure Estimates with Estimates Calculated Using EPA and Suggested Michigan Manufacturers Association Exposure Factors*. Princeton, N.J.: ENVIRON Corporation.

Gupta, Shreekant, George Van Houtven, and Maureen L. Cropper. 1993. *Do Benefits and Costs Matter in Environmental Regulation? An Analysis of EPA Decisions Under Superfund*. College Park: University of Maryland.

Hazardous Waste Cleanup Project. 1993. *Exaggerating Risk: How EPA's Risk Assessments Distort the Facts at Superfund Sites throughout the United States*. Washington, D.C.: Morgan, Lewrs, and Bockius.

Hird, John A. 1990. Superfund Expenditures and Cleanup Priorities: Distributive Politics or the Public Interest? *Journal of Policy Analysis and Management* 9 (4): 455–483.

Russell, Milton, E.W. Colglazier, and Mary R. English. 1991. *Hazardous Waste Remediation: The Task Ahead*. Knoxville: University of Tennessee, Waste Management Research and Education Institute.

U.S. Department of Commerce. 1992. *Statistical Abstract of the United States 1992*. Washington, D.C.: U.S. Government Printing Office.

U.S. EPA (Environmental Protection Agency). 1988. *CERCLA Compliance with Other Laws Manual: Interim Final*. Washington, D.C.: U.S. EPA, Office of Emergency and Remedial Response.

———. 1989a. *Risk Assessment Guidance for Superfund, Volume I: Human Health Evaluation Manual (Part A), Interim Final*. Washington, D.C.: U.S. EPA, Office of Emergency and Remedial Response.

———. 1989b. *Risk Assessment Guidance for Superfund: Volume I: Human Health Evaluation Manual (Part C, Risk Evaluation of Remedial Alternatives), Interim*. Publication 9285.7-01B. Washington, D.C.: U.S. EPA, Office of Emergency and Remedial Response.

———. 1991. *Role of the Baseline Risk Assessment in Superfund Remedy Selections*. Directive 9355 0–30. (April 22). Memo from Don Clay, Assistant Administrator, to directors of regional hazardous waste divisions. Washington, D.C.: U.S. EPA, Office of Solid Waste and Emergency Response.

———. 1993. *An SAB Report: Superfund Site Health Risk Assessment Guidelines*. EPA-SAB-EHC-93-007. Washington, D.C.: U.S. EPA, Science Advisory Board.

VERSAR, Inc. 1991. *Analysis of the Impact of Exposure Assumptions on Risk Assessment of Chemicals in the Environment*. Springfield, Virginia.

Viscusi, W. Kip. 1992. *Smoking: Making the Risky Decision*. Cambridge: Oxford University Press.

4

Do Benefits and Costs Matter in Environmental Regulation? An Analysis of EPA Decisions under Superfund

Shreekant Gupta, George Van Houtven, and Maureen L. Cropper

Of all environmental programs in the United States, the Superfund program is perhaps the most controversial. Under the program, the U.S. Environmental Protection Agency (EPA) is responsible for placing the most serious hazardous waste sites on the National Priorities List (NPL), for assessing the risks posed by the sites, and for selecting the most appropriate cleanup option for each site. By law, EPA is to choose a cleanup strategy that protects the health of people living near each site, regardless of cost. In practice, EPA has interpreted this requirement to mean that a lifetime risk of death from exposure to contaminants at the site should not exceed 1 in 10,000. Once this margin of safety is achieved, however, EPA is allowed to consider the cost of cleanup and to balance cost against other goals, such as the permanence of the cleanup and the acceptability of the cleanup option to the community.

One reason that the Superfund program is so controversial is that the cost of cleaning up sites is predicted to be high. EPA has estimated that the cost of cleanup will average $27 million per site (U.S. EPA 1990), implying that the cost of cleaning up all sites currently on the NPL—about 1,200 sites—could reach $30 billion. Although estimates vary widely, the cost of cleaning up both current and future sites on the NPL has been put at as much as $100 billion.[1]

ISSUES IN BALANCING BENEFITS AND COSTS AT SUPERFUND SITES

The high cost of cleanups has focused attention on the way in which EPA makes decisions about Superfund sites. One question that has been raised is whether the benefits of cleanup—which often accrue to a small population near the site—are worth the costs. This question is essentially a debate about the permanence of cleanup. At most sites, imminent danger of exposure to contaminants usually can be removed at low cost. Contaminated soil can be fenced off or capped, and an alternate water supply can be provided if groundwater is used for drinking. What raises the cost of cleanup is the decision to clean up the site for future generations: deciding, for instance, to incinerate contaminated soil, or to pump and treat an aquifer for thirty years to contain a plume of pollution.

To determine whether the benefits of long-term cleanup options are worth the cost is difficult. The first step to this end, which we take in this chapter, is to categorize cleanup strategies according to their permanence and to examine how much EPA has been willing to spend (or have others spend) for more permanent cleanups. For example, by estimating the value that EPA has implicitly attached to incinerating soil rather than placing it in a landfill, we raise the question: Is this the same value society would place on a more permanent cleanup?

Two Aspects of the Cleanup Decision

A second question that has been raised regarding cleanups is what criteria EPA uses to select a cleanup strategy for a site. One of the goals of this chapter is to see if the cleanup decisions that EPA has made at Superfund sites can be explained by a given set of factors. Specifically, we focus on two aspects of the decision to clean up contaminated soils at Superfund sites: deciding how much soil to excavate and/or cap, and choosing the treatment technology for the excavated soil.

The first decision—how much soil to remediate—affects current health risks to residents near a site. Typically, this decision is stated in terms of the concentration of contaminants above which all soil is excavated and/or capped. These concentrations are then mapped into a lifetime risk of death from exposure to hazardous substances at the site, which we term the *target risk level* for the site. It is this risk that is not to exceed 1 in 10,000.

If soil is to be excavated, EPA must decide how to dispose of it and whether disposal will occur on-site or off-site. This is essentially a decision about the permanence of cleanup. The least permanent cleanup is not to excavate soil at all, but to cap it. The cleanup, in this case, will last only as long as the life of the cap, and groundwater generally will not be protected from contamination. A more permanent solution is to excavate soil and put it in an approved landfill. This prevents exposure via groundwater (and other routes) as long as the landfill liner remains intact. An even more permanent solution (assuming pollutants are organic) is to incinerate the soil.

EPA Cleanup Criteria

Our goal is to see what factors explain each aspect of the cleanup decision. Specifically, we wish to see whether EPA has considered costs in choosing target risk levels (even though it is not supposed to do so by law) and whether it has considered costs in choosing the permanence of the cleanup option.

It is also natural to ask whether the benefits of lower target risk levels or more permanent cleanup options have influenced cleanup decisions. This question is, however, difficult to address since EPA does not measure the additional number of lives (or life-years) saved by different cleanup strategies. What we *can* ask is whether variables that are likely to be correlated with the benefits of cleanups—such as the size of the population living near the site—can explain the decisions made. It should also be kept in mind that life-years saved are not necessarily the only benefits of the Superfund program.

Environmental Equity

Finally, we wish to shed some light on an issue that has received much attention in the last several years, but little careful study—the issue of environmental equity and, in particular, of risk equity. During the past several years, environmental and other advocacy groups have charged that minorities and the poor suffer disproportionately from the effects of pollution (United Church of Christ 1987). In the case of Superfund, it has been alleged (Lavelle and Coyle 1992) that EPA selects less complete and less permanent cleanups in areas that have a high percentage of poor and/or minority residents. These allegations, however, are based on simple correlations of race and permanence that fail to hold other factors constant. We wish to see, holding other factors constant, whether EPA, in fact, has selected higher target risk levels or less permanent cleanups in areas that have a high percentage of minority residents or low median household incomes.

Addressing the Issues: Data and Models

To shed light on these issues, we have gathered data on the decisions to clean up contaminated soils at 110 Superfund sites—all wood-preserving sites for which cleanup decisions had been made as of the end of fiscal year 1991 and selected sites with PCB (polychlorinated biphenyl) contamination in excess of 10 parts per million (ppm). We have used the data to model the decision to clean up contaminated soils at these sites.[2] This model, which is presented formally in the appendix to this chapter, leads to two equations to be estimated.

The first equation explains the target risk level at each site as a function of baseline health risks at the site, two measures of cleanup costs—the size of the site and the cost of excavating and treating a cubic yard of soil at the site—and other site characteristics. These include whether the site is located in an urban area, the percentage of nonwhite residents living in the zip code in which the site is located, and median household income for the zip code. Also included in this equation is a dummy variable to indicate whether potentially responsible parties (PRPs) were in charge of the remedial investigation/feasibility study.

The second equation is a discrete choice model that explains which cleanup option EPA chose at a site, as a function of the cost of each option, whether the option did or did not involve excavation of soil, whether contaminated soil was treated, and whether treatment occurred on-site or off-site. The characteristics of cleanup (excavate/don't excavate, on-site/off-site, and so forth) are also interacted with site characteristics (for instance, percentage of population that is nonwhite) to see if these influenced the nature of the cleanup selected.

This chapter covers the issues that are central to cleanup decisions, beginning with a discussion of the model describing the cleanup decision. (The equations and other formal aspects of this model are given at the end of the chapter.) The next section describes the data collected, while the subsequent section contains empirical estimates of the target risk equation. A detailed examination of the discrete choice of cleanup option follows, and a final section summarizes and draws some conclusions from our work.

A MODEL OF THE CLEANUP DECISION

At each Superfund site with contaminated soils, EPA must decide how extensively to clean up the site (how much soil to excavate or cap), as well as how permanent the cleanup will be. As noted earlier, the extent of cleanup is a continuous variable that measures the maximum allowable concentration of a hazardous substance remaining at a site after cleanup. (Usually, there will be a vector of target concentrations, one for each of the major contaminants at the site.) In general, there is a direct relationship between this *target concentration* and the lifetime risk of death associated with the site, or *target risk*. The permanence decision corresponds to choosing the technology that will be used for cleanup (capping, soil flushing, incineration) from a set of several alternative technologies.

A formal model of the choice of technology and target concentration is presented in the appendix to this chapter. In nontechnical terms, the model views the decisionmaker as choosing an optimal concentration above which all soil will be excavated or capped, such as 10 ppm of PCBs. It is reasonable that this concentration will vary with the cleanup technology considered. For example, a lower target concentration might be chosen if excavated soil were to be put in a landfill than if the soil were to be incinerated. The model assumes that an optimal target concentration is chosen for each cleanup technology considered at the site; then, the appropriate cleanup technology is chosen.

For each technological option considered, the model assumes that the target concentration chosen (or, equivalently, the target risk level chosen) balances benefits against costs. For each technological option, lowering the target risk level will raise the number of lives saved from the cleanup but also raise the cost. The model assumes that a target risk level is chosen that maximizes benefits—defined primarily in terms of lives saved—minus costs.

Other things being equal, one would expect that a lower target risk level would be chosen at sites where the cost of excavating a cubic yard of soil is low. One might also expect a lower target risk level at smaller sites than at larger sites, since the former are cheaper to clean up. The life-saving benefits of a low target risk level are greater at sites surrounded by a large population than at sites surrounded by a small population; hence one would expect the target risk level to vary inversely with the exposed population living near a site.

In addition to these factors, the model predicts that other characteristics of a site may influence the target risk level chosen. These include characteristics of the population living near the site, which may affect the value the decisionmaker puts on the life-saving benefits associated with cleanup. The initial risk of death associated with the site (excess lifetime cancer risk) may also affect the value attached to a given reduction in risk of death by the decisionmaker.

Once an optimal target risk level has been chosen for each technological option considered, the decisionmaker is assumed to choose the technological option (landfilling of waste, incineration, and so forth) that yields the highest net benefits. The benefits of each technological option are characterized by the permanence of the cleanup (for example, incineration of waste is more permanent than capping it) and by whether the cleanup occurs on-site or off-site. The characteristics of the site may also affect the value attached to more permanent cleanups or to off-site cleanups. For example, EPA may attach more importance to off-site cleanups at sites located in high-income neighborhoods than at sites located in low-income neighborhoods.

Other things being equal, one would expect that technological options yielding high benefits (such as more permanent options) would be more likely to be selected than options yielding low benefits. Likewise, options with lower costs would be more likely to be selected than options with higher costs.

THE DATA

In selecting data to estimate the model of cleanup decisions, we were limited to those NPL sites for which records of decision (RODs)—documents describing the cleanup strategy chosen by EPA—had been signed. Of the 945 sites for which RODs had been signed as of the end of fiscal year 1991, we selected 110 sites: 32 wood-preserving sites and 78 sites with PCB contamination.[3] There is a total of 127 RODs for the 110 sites, since a single site may have more than one operable unit, a portion of the site that is treated separately for purposes of cleanup. The RODs sometimes contain inaccuracies but are the only readily available source of information about the remedy chosen for a site.

Wood-preserving sites are wood treatment facilities where pentachlorophenol (PCP) or creosote was used to pressure-treat wood to prevent it from rotting. Soils at these sites are contaminated with polyaromatic hydrocarbons (PAHs), constituents of creosote that are considered probable human carcinogens. The PCB sites in the sample include landfills, former manufacturing facilities, and other sites where PCBs—also considered probable human carcinogens—are found.[4]

These sites were selected for two reasons. First, because their principal contaminants are carcinogenic, estimates of health risks from each site are more likely to be available than for sites whose contaminants are not carcinogenic. Second, because both sets of sites contain organic pollutants, the technological options available for cleaning up contaminated soils are similar at both sets of sites.

For each site (more accurately, for each operable unit), data were gathered from the ROD on the set of cleanup options considered and on the characteristics of the site. The model discussed in the previous section suggests that, for each cleanup option considered, we would like to know the cost of the option, the target risk level (or concentration) associated with the option, and the permanence of the option. While data on the cost of each option are available, the target risk level is reported only for the cleanup option actually selected by EPA. The permanence of each option is likewise not reported in the ROD; however, we have developed a scheme to characterize the permanence of each cleanup option. (This scheme is described later in this chapter in the section on the role of benefits and costs in the choice of options.)

Site-specific variables, which may influence both the target risk level and the permanence of the cleanup option chosen, include the state and EPA region in which the site is located, the size of the site, whether the site is located in an urban or rural area, and the population living either within a one-mile or a three-mile radius of the site. Also included in the site description are the baseline cancer risk associated with the site and the site's Hazard Ranking System score.

WHY DO TARGET RISKS VARY ACROSS SUPERFUND SITES?

In the RODs, the soil cleanup goal is quantified in terms of a target risk level (the postcleanup lifetime cancer risk attributable to a site) and/or as target concentrations of pollutants at the site. The advantage of focusing on target risk rather than target concentration is that the former collapses all contaminants into a common metric. Unfortunately, target risk levels are reported for only 61 of the 127 decisions in our database. Furthermore, as noted above, target risk levels are not reported for all cleanup options considered, only for the option chosen. This means that our analysis is based on only 61 observations.

The earlier discussion (in the section on the model of the cleanup decision) implies that the target risk level should depend on the population living near the site, the baseline risk associated with the site, the size of the site, unit treatment costs, and site characteristics. The means and standard deviations of these variables, for both wood preserving sites and PCB sites, appear in Table 1. The target risk level ranges from 10^{-4} to 10^{-7}. Thus, while EPA's guidance suggests a target risk level between 10^{-4} and 10^{-6}, the guidance is not strictly adhered to.

Variables That Should Affect Target Risk

To the variables that should affect target risk according to the model discussed earlier, we have added two variables: a time trend to reflect a

Table 1. Means and Standard Deviations of Variables in Target Risk Equation

Variable	Wood-preserving sites			PCB sites		
	N	Mean	Standard deviation	N	Mean	Standard deviation
Target risk level	24	8.96E–06	2.05E–05	37	2.46E–05	4.79E–05
Time trend (1983 = 1)	24	6.92	1.1	37	6.84	1.59
Whether PRP-lead RI/FS	24	0.42	0.5	37	0.51	0.51
Whether urban setting	24	0.21	0.41	37	0.27	0.45
Baseline future risk	17	2.90E–01	0.44	29	1.16E–01	0.18
Groundwater HRS score	24	60.02	16.24	37	59.77	21.78
Surface water HRS score	24	14.29	10.68	37	23.57	19.51
Area (acres)	24	59.73	56.62	37	149.22	587.37
Unit treatment cost (1987 $)	24	348.92	365.15	37	589.98	579.79
Percent nonwhite population	24	20.22	20.7	37	13.45	26.1
Median income (1989 $)	24	26,233	14,168	37	28,047	11,184

Notes: N = number of sites. The following convention is used in this table: E–06 = 1/1,000,000 (for example, 8.96E–06 = 0.00000896); E–05 = 1/100,000 (for example, 2.05E–05 = 0.0000205); and so forth.

possible trend in the thoroughness of cleanups over time and a variable to indicate who was in charge of assessing the risks and cleanup options at the site (the "lead" on the Remedial Investigation/Feasibility Study, or RI/FS). Although the regional EPA administrator is ultimately responsible for selecting a cleanup strategy for a site, the RI/FS—the detailed site study—that precedes the choice of cleanup strategy may be conducted either by EPA (at a "fund-lead" site) or by the people responsible for cleaning up the site (the potentially responsible parties, or PRPs) at a PRP-lead site. It is sometimes thought that the party responsible for the site investigation can influence the menu of alternatives considered for cleanup and, hence, the cleanup goal selected at the site.

Among the variables suggested by the model that should explain the target risk selected, we have substituted for the size of the exposed population—a variable with many missing observations—a dummy variable indicating whether the site is located in an urban area. (This is an admittedly crude proxy for population density near the site.) If the urban dummy does reflect the size of the exposed population, it should be negatively related to target risk, since the higher the exposed population, the greater are the benefits of lowering the target risk level.

We have included baseline risk of death from exposure to the site as a possible determinant of target risk level. If utility is a linear function of lives saved, then there is no reason that baseline risk should matter: all risk reductions are equally valuable, regardless of the baseline. If, however, the utility received from saving an additional life diminishes with the number of lives saved, baseline risk will matter.[5]

Several measures of baseline risk are available. Baseline future risk measures risk of cancer to the "maximally exposed individual" from all exposure pathways, assuming nothing is done to clean up the site.[6] This measure may be disaggregated into risk attributable to direct contact with contaminated soil and risk attributable to exposure to contaminated groundwater.[7]

Two features of baseline risk stand out. First, the risk of cancer at the sites studied is estimated to come primarily from contaminated groundwater, rather than from direct contact with contaminated soil. Second, the magnitude of the lifetime cancer risks from these sites—which is huge by contrast with most environmental risks—reflects the conservative assumptions used to estimate exposure.

Estimates of baseline future risk are based on a quantitative risk assessment conducted once a site has been placed on the NPL. To determine which sites will be placed on the NPL, a preliminary scoring of hazards is conducted using the Hazard Ranking System. The Hazard Ranking System (HRS) score, which is computed separately for groundwater and for surface water, is a quick-and-dirty estimate of the potential for exposure to hazardous substances at a site. We are interested in seeing whether this rough-and-ready estimate of hazard potential has as much influence on the choice of target risk level as do formal risk assessments.

The effect of cost considerations on choice of target risk is captured by the area and unit treatment cost variables. The surface area of contamination is an exogenous measure of the size of the site.[8] If costs matter, we would expect, other things being equal, that higher target risks would be chosen at larger sites and at sites with higher unit treatment costs.

The last two variables in Table 1 describe characteristics of the site. To see whether less thorough cleanups (higher target risk levels) are chosen in poor areas or in areas with a high percentage of nonwhites, the variables "Percent nonwhite population" and "Median income" are included in the equation. Both variables are measured for the zip code in which the Superfund site is located and are based on 1990 U.S. Census data.

Empirical Results

Estimates of the target risk equation are presented in Table 2 (for wood-preserving sites) and Table 3 (for PCB sites). The dependent variable is the natural logarithm of the target risk level. In Specification 1, baseline future risk is included as an explanatory variable, whereas Specification 2 replaces this with the groundwater and surface water HRS scores. Specifications 1A and 2A exclude the variables "Percent nonwhite pop-

Table 2. Factors Affecting Target Risk Levels for 24 Wood-Preserving Sites

Explanatory variables	Specification			
	1A	1B	2A	2B
Intercept	−11.69	−10.19	−9.96	−9.62
	(−5.25)	(−5.07)	(−4.62)	(−3.64)
Time trend (1983 = 1)	−0.13	−0.23	−0.22	−0.2
	(−0.41)	(−0.72)	(−0.93)	(−0.94)
PRP-lead	0.63	0.78	1.04	1.38
	(1.18)	(1.62)	(2.0)	(2.84)
Urban setting	1.94	3.07	2.19	3.83
	(3.57)	(3.42)	(3.39)	(3.81)
Baseline future risk	−0.49	−0.22	−9.37E−03	7.36E−03
× (1 − future risk missing)	(−1.0)	(−0.45)	(−0.54)	(0.33)
Future risk missing	−0.27	−0.65	−3.78 E−02	−4.69E−02
	(−0.56)	(−1.05)	(−2.35)	(−2.54)
Area	−3.6E−03	−7.73E−03	−7.45E−03	−1.12E−02
	(−0.73)	(−1.81)	(−1.86)	(−2.55)
Unit treatment cost	−1.5E−03	−2.15E−03	−1.85E−03	−2.02E−03
	(−2.89)	(−2.94)	(−2.17)	(−2.82)
Percent nonwhite population		−2.85E−02		−4.65E−02
		(−1.40)		(−1.93)
Median income		2.31E−08		−2.412E−05
		(0.001)		(−1.29)
R-squared:	0.49	0.55	0.53	0.63
Breusch-Pagan statistic:	6.02	12.57	3.12	10.78
Degrees of freedom:	16	14	16	14

Notes: The dependent variable is the natural logarithm of the target risk level. In Specification 1, baseline future risk is included as an explanatory variable, whereas Specification 2 replaces this with the groundwater and surface water Hazard Ranking System (HRS) scores. Specifications 1A and 2A exclude the percent nonwhite population and median income variables from the equation, whereas both variables are included in the 1B and 2B versions of the equations. Dark gray shading indicates the groundwater HRS score; light gray shading indicates the surface water HRS score.

T-ratios are in parentheses and are corrected for heteroscedasticity. The following convention is used in this table: E−06 = 1/1,000,000 (for example, 8.96E−06 = 0.00000896); E−05 = 1/100,000 (for example, 2.05E−05 = 0.0000205); and so forth.

ulation" and "Median income" from the equation, whereas both variables are included in the 1B and 2B versions of the equations.

Three sets of results stand out regarding the determinants of target risk levels. First, costs do not seem to have been considered in choosing a target risk level, either at wood-preserving or PCB sites. Second, the concern that EPA has discriminated against minorities by choosing less thorough cleanups in minority areas seems largely unfounded. Third, EPA has shown a clear preference for more thorough cleanups at sites with high HRS scores.

If EPA considered costs in setting target risk levels, one would expect the coefficients of "Area" and "Unit treatment cost" to be positive and significant. At PCB sites, both variables are insignificant. At wood-

Table 3. Factors Affecting Target Risk Levels for 37 PCB Sites

Explanatory variables	Specification			
	1A	1B	2A	2B
Intercept	−14.19	−13.29	−15.31	−14.99
	(−13.25)	(−6.65)	(−10.35)	(−8.56)
Time trend (1983 = 1)	0.37	0.38	0.47	0.47
	(2.14)	(2.06)	(2.72)	(2.67)
PRP-lead	−0.31	−0.41	1.1E−02	6.88E−03
	(−0.31)	(−0.39)	(1.1E−02)	(1.0E−02)
Urban setting	0.58	0.87	0.59	0.66
	(0.80)	(0.9)	(0.54)	(0.5)
Baseline future risk	−3.6	−3.81	−5.70E−04	1.06E−03
× (1 − future risk missing)	(−1.14)	(−1.16)	(−4.0E−02)	(7.0E−02)
Future risk missing	−1.44	−1.7	−7.40E−04	−2.24E−03
	(−2.66)	(−2.20)	(−0.03)	(−0.11)
Area	3.53E−04	1.14E−04	5.60E−04	5.00E−04
	(0.4)	(0.13)	(0.57)	(0.52)
Unit treatment cost	−2.81E−04	−4.44E−05	−8.30E−04	−7.56E−04
	(−0.30)	(−5.0E−02)	(−0.77)	(−0.73)
Percent nonwhite population		−1.41E−02		−5.38E−03
		(−0.75)		(−0.31)
Median income		−2.886E−05		−1.488E−05
		(−0.62)		(−0.33)
R-squared:	0.2	0.22	0.14	0.14
Breusch-Pagan statistic:	19.72	21.53	18.24	21.44
Degrees of freedom:	29	27	29	27

Notes: The dependent variable is the log of the target risk level. In Specification 1, baseline future risk is included as an explanatory variable, whereas Specification 2 replaces this with the groundwater and surface water Hazard Ranking System (HRS) scores. Specifications 1A and 2A exclude the percent nonwhite population and median income variables from the equation, whereas both variables are included in the 1B and 2B versions of the equations. Dark gray shading indicates the groundwater HRS score; light gray shading indicates the surface water HRS score.

T-ratios are in parentheses and are corrected for heteroscedasticity. The following convention is used in this table: E−06 = 1/1,000,000 (for example, 8.96E−06 = 0.00000896); E−05 = 1/100,000 (for example, 2.05E−05 = 0.0000205); and so forth.

preserving sites, the sign of "Unit treatment cost" is negative, suggesting that more complete cleanups have been selected at sites with higher unit costs. The "Area" coefficient is significant only in Specifications 2A and 2B in Table 2; however, its sign is also negative, which is the opposite of what theory suggests. As noted in the opening section of this chapter, EPA is not supposed to consider costs in setting target risk goals, and it seems to have obeyed this mandate.

As far as risk equity is concerned, both the percentage of population that is nonwhite and the median income of the zip code in which a site is located have no effect on the choice of target risk at PCB sites. The percentage of the population that is nonwhite is significant in one of the equations for wood-preserving sites; however, it has a negative coefficient, implying that target risks are *lower* in minority areas. Based on this small sample of sites, concern that minorities are discriminated against in choosing cleanup goals thus appears unfounded. Median income is never a statistically significant determinant of target risks.

The third result that holds for both sets of sites is the importance of baseline risk in selecting a target risk level. While baseline future risk per se is not significant in explaining variation in target risks, the HRS surface water score is significant at wood-preserving sites: sites with higher HRS scores are cleaned up more thoroughly (have lower target risks) than sites with lower HRS scores. One possible explanation for this result is that EPA, in assigning cleanup priorities to sites, ranks sites on the basis of HRS score. Even this practice, however, does not fully explain why dirtier sites would be cleaned up more thoroughly: while one would expect a given risk *reduction* to be valued more highly when the baseline risk is higher, it is hard to explain why the target risk *level* selected would be lower.

Another result that is hard to explain is that higher target risks are selected in urban areas than in rural areas, even though the former are presumably more densely populated. This anomaly may occur because more stringent cleanups are usually accompanied by a greater amount of soil excavation: officials are reluctant to excavate soils in more densely populated areas, since this exposes many people to short-term risks. Another possible explanation is that the urban dummy acts as a proxy for industrial land use, which diminishes concern for health risks.

Finally, it is worth noting that, at wood-preserving sites, less thorough cleanups were selected when PRPs were in charge of the RI/FS than when EPA or a state government was. This supports the hypothesis that the party in charge of the RI/FS may influence the cleanup goal chosen. As far as time trends are concerned, PCB sites apparently have been receiving less thorough cleanups over time.

THE ROLE OF BENEFITS AND COSTS IN CHOICE OF CLEANUP TECHNOLOGY

We now focus on the other aspect of the cleanup decision: the choice of which technology to use in cleaning up the site. In practice, each ROD lists from three to twelve options that are considered for cleaning up the site. Each option consists of a description of the technology used and the amount of soil addressed. As noted earlier in the discussion of the model of the cleanup decision, the choice of technology to use in cleaning up a site should be made by comparing the net benefits of each cleanup option. The net benefits of the option are a function of the number of lives saved, the permanence of the option, and its cost. Of these three components, only the cost of the option is noted in the ROD.

A Classification Scheme for Cleanup Options

To describe the permanence of cleanup options, we developed a classification scheme that is based on two aspects of each alternative: whether the alternative involves excavation of contaminated soil and whether the alternative involves treatment of the contaminated soil. In addition, we distinguished whether remedies that entail excavation are conducted on-site or off-site. Combining these choices yields a total of six categories of remedial alternatives:
- On-site treatment of soil that has been excavated (on-site treatment)
- Off-site treatment of soil that has been excavated (off-site treatment)
- Disposal of excavated but untreated soil in a landfill at the site (on-site landfill)
- Disposal of excavated but untreated soil in a landfill off the site (off-site landfill)
- Containment of soil that has been neither excavated nor treated (containment)
- On-site treatment of soil that has not been excavated (in situ treatment)[9]

The six categories are diagrammed in Figure 1.

Table 4 lists, for both wood-preserving and PCB sites, the number of times each category was considered and selected, and the unit cost of cleanup options within each category. All six categories may not be considered at a site, whereas some, such as on-site treatment, may be considered more than once. Of the six categories, on-site and off-site treatment correspond to the most permanent cleanups. According to the 1986 amendments to the Superfund law—the Superfund Amendments and Reauthorization Act (SARA)—EPA is supposed to show a preference for treatment, as opposed to nontreatment, alternatives. We also have distinguished whether disposal and/or treatment of excavated soil occurred on- or off-site because of the controversy surrounding off-site cleanups. Off-site cleanups are often favored by people living near a Superfund site, since this type of cleanup is perceived to be a permanent solution to the problem. The SARA amendments, however, indicate a preference for on-site, as opposed to off-site, remedies. We wish to see whether EPA has, in fact, exhibited such a preference.

Table 4 illustrates the magnitude of the permanence-cost trade-off facing environmental officials. The average cost of the least permanent options—containment and on-site landfill—is approximately one order of magnitude smaller than the average cost of on-site treatment. Never-

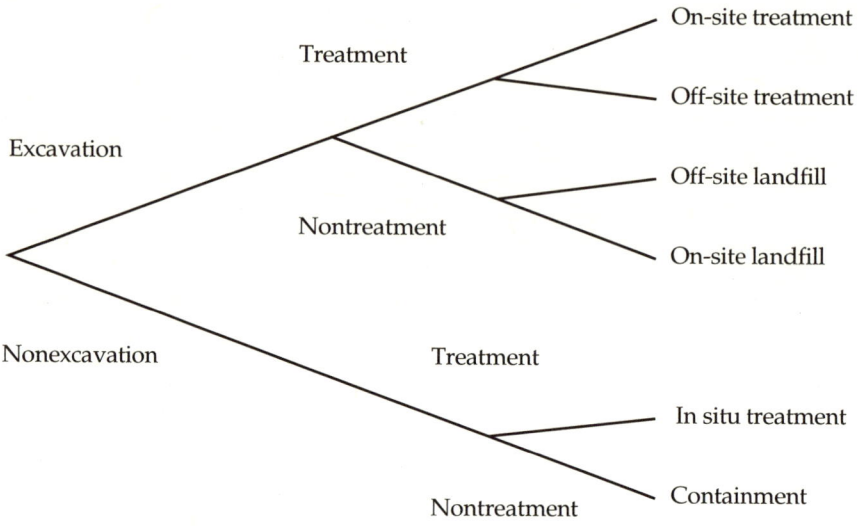

Figure 1. Remedial Alternatives for Soil Contamination

theless, on-site treatment was the most preferred of the six cleanup categories: it was selected 73% of the time at wood-preserving sites and 62% of the time at PCB sites. For this reason, on-site treatment has been further broken down into three subcategories–incineration, innovative treatment, and solidification/stabilization.

The Choice of Technology at Wood-Preserving Sites

Separate equations were estimated to explain the remedial alternatives selected at wood-preserving sites and at PCB sites. In examining these results, we focus on three questions:
- Did costs matter to EPA in its choice of cleanup option? That is, was the agency more likely to select an inexpensive cleanup than an expensive one, other things equal?
- Did EPA show a preference for more permanent cleanups, and, if so, how much was it willing to pay for them?
- Did EPA's propensity to select one option rather than another vary with site characteristics?

Table 5 presents the model for wood-preserving sites. Two results stand out. First, EPA is less likely to choose a cleanup option the more costly it is. Costs *do* matter in determining which technology to use in cleaning up a

Table 4. Cleanup Options Considered and Selected and Their Average Costs

	Mean cost [a]		Wood-preserving sites				PCB sites		
	Per unit ($/cubic yard)	Total ($ million)	N	Mean volume (cubic yards)	Standard deviation of volume		N	Mean volume (cubic yards)	Standard deviation of volume
Remedial options considered									
Excavation Alternatives									
On-site landfill	144	6.1	16	36,053	28,754		29	45,877	59,593
Off-site landfill	619	7.9	15	18,136	14,692		50	77,058	224,229
Off-site treatment	1,428	45.5	19	38,351	37,896		33	26,235	61,115
On-site treatment	350	13.1	85	44,881	48,097		156	55,555	141,736
Incineration	555	22.0	29	40,639	38,508		67	53,577	110,364
Innovative treatment	252	9.7	45	42,826	38,281		58	44,535	50,326
Stabilization/solidification	211	3.9	11	20,038	21,282		31	80,450	267,022
Nonexcavation Alternatives									
In situ treatment	232	11.3	12	42,262	38,312		11	45,810	38,003
Containment	79	3.5	23	46,549	46,355		36	128,850	282,599
All sites	430	14.2	170	41,536	43,030		315	63,042	167,760
Remedial options selected									
Excavation Alternatives									
On-site landfill	67	3.4	2	34,875	15,380		6	42,050	69,324
Off-site landfill	763	4.8	3	14,651	20,118		13	9,079	10,110
Off-site treatment	655	17.5	1	26,733	–		4	534	446
On-site treatment	329	10.9	29	36,529	45,624		54	32,905	31,982
Incineration	486	21.2	8	39,627	34,610		22	34,298	33,103
Innovative treatment	267	8.0	16	32,127	33,628		18	32,295	30,903
Stabilization/solidification	279	3.7	5	11,924	6,598		14	31,501	33,841
Nonexcavation Alternatives									
In situ treatment	142	7.6	2	66,150	62,013		1	149,000	–
Containment	31	0.4	3	35,733	42,287		9	421,222	467,160
All sites	325	9.3	40	36,856	42,920		87	69,993	189,503

Note: N = number of sites. [a] The cost figures refer to wood-preserving sites only and are in 1987 prices.

Table 5. Choice of Remedial Action at Wood-Preserving Sites

Cost and technology variables	Specification 1	Specification 2	Specification 3	Specification 4: Time trend	Specification 5: Race	Specification 6: Income
Log cost (1987 $)	−0.694	−0.909	−0.699	−3.741	−0.778	0.001
	(−2.39)	(−2.50)	(−2.43)	(−2.39)	(−1.47)	(0.001)
Log cost × site variable				0.497	−0.002	−0.000027
				(2.08)	(−0.11)	(−0.72)
On-site landfill	0.536	0.749	0.537	5.125	2.447	1.531
	(0.51)	(0.68)	(0.51)	(1.27)	(1.18)	(0.43)
Off-site landfill	1.291	1.701				
	(1.08)	(1.32)	**1.244**	**5.511**	**−1.045**	**2.656**
Off-site treatment	1.133	1.627	**(1.09)**	**(1.31)**	**(−0.48)**	**(0.62)**
	(0.77)	(1.05)				
On-site excavation and treatment	**2.290**		**2.301**	**7.213**	**2.264**	**3.694**
	(2.48)		**(2.50)**	**(1.73)**	**(1.31)**	**(1.06)**
Incineration		3.126				
		(2.48)				
Solidification/ stabilization		2.523				
		(2.20)				
Innovative treatment		2.419				
		(2.44)				
In situ treatment	1.088	1.306	1.096	5.306	0.527	5.448
	(0.93)	(1.08)	(0.94)	(0.70)	(0.25)	(1.24)
Secondary treatment	1.380	1.551	1.389	5.445	−0.483	0.961
	(1.97)	(2.11)	(1.99)	(1.81)	(−0.4)	(0.48)
On-site landfill × site variables				−1.0	−0.315	−2.30E−05
				(−0.96)	(−0.95)	(−0.22)
Off-site remedies × site variables				−0.722	0.134	−0.000034
				(−1.11)	(1.32)	(−0.25)
On-site excavation and treatment × site variables				−0.752	0.014	−0.000034
				(−1.24)	(0.17)	(−0.30)
In situ treatment × site variables				−0.645	0.04	−0.000186
				(−0.57)	(0.48)	(−0.89)
Secondary treatment × site variables				−0.576	0.103	1.19E−05
				(−1.32)	(1.83)	(0.17)
Log likelihood:	−44.12	−43.56	−44.12	−39.91	−39.48	−42.69

Notes: Specification 1 estimates the choice of cleanup option expressed as a function of cost and technology dummies. Specification 2 is Specification 1 disaggregating on-site excavation and treatment into three categories. Specification 3 aggregates the off-site dummy variables. Specifications 4, 5, and 6 interact site variables (time trend, race, income, respectively) with the log cost and technology dummies. For Specification 4, "Time trend" signifies the year the ROD was signed, with 1983 = 1. For Specification 5, Race signifies the percentage of the nonwhite population in the zip code where the site is located. For Specification 6, Income signifies the median household income in the zip code where the site is located.

T-ratios are in parentheses. Coefficients in boldface represent aggregated categories.

wood-preserving site. Second, EPA has demonstrated a clear preference for on-site excavation and treatment at wood-preserving sites.

Both results appear clearly in the first column (Specification 1) of Table 5, which explains the choice of cleanup option solely as a function of cost and of the technology dummies. The logarithm of cost is significant and negative, indicating that the higher the cost of a cleanup option, the less likely it is to be chosen. Of the five technology dummies described above (containment is the omitted category), only on-site excavation and treatment is statistically significant. This implies that EPA was willing to pay significantly more for on-site excavation and treatment, the most permanent technology, as compared to capping; however, it was willing to pay no more for the other four categories in Figure 1 than for capping. The variable "Secondary treatment" indicates that, in addition to the method of treatment applied to the majority of waste at the site (which is characterized by the technology dummies), an additional method was used to treat a portion of waste at the site. The significance of this variable indicates EPA's preference for using more than one treatment technology at a site.

Specifications 2 and 3 of the table present, respectively, a more detailed and a less detailed characterization of cleanup options. Specification 2 disaggregates on-site excavation and treatment into three categories—incineration, solidification, and innovative treatment. While each of the three categories is statistically significant—EPA was willing to pay a premium for any one of them relative to capping—their coefficients are not significantly different from one another. A comparison of Specifications 1 and 3 likewise indicates that the coefficients for the two off-site options are not significantly different from one another.

The remainder of Table 5 interacts site characteristics and a time trend with log cost and with the technology dummies. Only two such interactions are significant: secondary treatment (the use of more than one treatment technology) was more likely to be used the higher the percentage of minority residents near the site, and costs mattered less in remedy selection over time. We emphasize, however, that for the sites studied there is no evidence in Table 5 that EPA selected less permanent remedies in areas with a large minority population or in low-income areas. All interactions between the permanence (technology) dummies and either race or income are insignificant.

The second result—that costs matter less over time—accords with the spirit of the SARA amendments; that is, EPA should give more weight to permanent remedies rather than to costs in choosing a cleanup option. However, a strict test of the amendments—interacting the post-SARA dummy with on-site excavation and treatment—does not yield significant results.

One of the implications of Table 5 and of alternate specifications not reported in the table is that the weight attached to cost and to the technology dummies seems to vary little with site characteristics: EPA's propensity to choose one cleanup option over another was consistent across sites. In particular, it was unaffected by whether the site was located in an urban area, by baseline risk, or by risk of groundwater contamination.

The Value of More Permanent Cleanup Options

Since costs and permanence are both statistically significant in explaining the cleanup option chosen, one can compute the rate at which EPA was willing to substitute cost for permanence to determine an implicit willingness to pay (or have polluters pay) for increased permanence. Formally, one can ask how much costs can be increased while changing the cleanup option from containment to on-site excavation and treatment, and keep net benefits constant.

Specification 1 of Table 5 implies that, at a site where capping would cost $400,000 (1987 dollars), EPA would be willing to spend an additional $11.4 million to incinerate the soil. Its willingness to pay for on-site innovative treatment or stabilization (over the cost of capping) was about half as much ($5 million and $5.7 million, respectively).

It is important to emphasize what these implicit valuations measure. The $11.4 million value attached to incineration is not simply the difference in cost between on-site incineration and containment (capping) at sites where incineration was chosen. Indeed, this cost difference—$21.2 million versus $0.4 million (see Table 4)—is greater than the valuation of $11.4 million implied by Table 5. What Table 5 reflects is that EPA sometimes chose not to incinerate soil, even when it was relatively inexpensive to do so. This lowers the implicit valuation of the option below average cost at sites where incineration was chosen.

The Choice of Technology at PCB Sites

Table 6 presents models of the choice of cleanup options at PCB sites. At PCB sites, costs clearly played a role in the selection of cleanup technology: as reflected in the specification columns of Table 6, more expensive technologies are less likely to be selected, other things being equal. The disutility attached to cost, however, was less at larger sites (up to 15,000 cubic yards) than at smaller sites.[10] Costs in general tend to rise with the amount of contaminated material, and EPA appears to be less averse to additional costs at larger sites. If the benefits of site remediation increase with the volume of contaminated material present, this is a desirable outcome. However, a larger volume of contaminated soil at a site does not necessarily mean that it is more of a threat. Table 6 also suggests that EPA was willing to pay more for more permanent cleanups at PCB sites. Of all the categories in Figure 1, on-site treatment (in practice, on-site incineration) is clearly the most valuable: its coefficient exceeds those of the other technology dummies in all the specifications in Table 6.[11] In fact, Specification 1 of the table implies that EPA was willing to pay $33.5 million (1987 dollars) more for on-site treatment than it was willing to pay to contain the waste or treat it in situ.[12] Off-site treatment (in practice, off-site incineration) was nearly as valuable as on-site treatment. It is the second most preferred technology in all equations in the table, and commands a value in Specification 1 of $22.3 million, relative to nonexcavation cleanups. The fact that off-site treatment is somewhat less valuable than on-site treatment reflects the fact that it was chosen less often than on-site treatment, which accords with the spirit of the SARA amendments.

It is not surprising that EPA was willing to pay more for the on-site and off-site treatment alternatives than for other cleanups: excavation and treatment (usually incineration) of contaminated soil is the most permanent method of disposing of PCBs. What is, perhaps, surprising is that disposing of waste in an off-site landfill—a less permanent alternative—was valued about as highly as off-site incineration. The value of an off-site landfill (relative to nonexcavation) is $25.3 million in Specification 1—approximately the same value as off-site treatment. Indeed, the hypothesis that the two cleanup options have identical coefficients—compare Specifications 3 and 4—cannot be rejected. A plausible explanation for this similarity in values is that EPA's preferences reflect the preferences of local residents, who view all cleanups that remove waste from the site as equally permanent.

Off-site landfills are clearly valued more highly than on-site landfills. The latter category is valued no more highly than nonexcavation cleanups in Specifications 1 and 2.

The Effect of Site Characteristics on Choice of Technology

In Specifications 3 through 10 of Table 6, the values attached to treatment and to off-site disposal are allowed to vary with volume of waste at the site. In all cases, the value attached to treatment or to a landfill decreases with the size of the site. A possible rationale for this finding is that, at large sites, excavation of soil will expose more people to short-term hazards than at small sites. Cleanup options involving excavation are therefore less attractive at large sites than at small sites.

When volume of waste is interacted with the technology dummies, on-site treatment still remains the most preferred of the six cleanup tech-

Table 6. Choice of Remedial Action at PCB Sites

Cost and technology variables	Specification 1	Specification 2	Specification 3	Specification 4	Specification 5	Specification 6	Specification 7: Hazard Ranking System score[a]	Specification 8: Race	Specification 9: Median household income[b]	Specification 10: Per capita income[b]
Cost (millions of 1987 $)	-0.08 (-3.73)	-4.07 (-3.09)	-3.49 (-2.18)	-3.48 (-2.60)	-3.41 (-2.55)	-3.50 (-2.62)	-3.44 (-2.55)	-4.28 (-2.60)	-3.43 (-2.54)	-3.40 (-2.51)
Cost × Lvol1[c]		0.42 (3.03)	0.36 (2.14)	0.36 (2.56)	0.35 (2.48)	0.34 (2.46)	0.35 (2.49)	0.44 (2.56)	0.35 (2.49)	0.35 (2.45)
Cost × current soil risk[d]					0.93 (1.56)					
Cost × current soil risk missing dummy variable					0.01 (0.17)					
Cost × Hazard Ranking System score[a]						0.002 (1.58)				
Off-site				32.88 (2.56)	32.30 (2.47)	37.73 (2.74)	40.34 (2.60)	30.57 (2.36)	36.49 (2.14)	37.40 (1.96)
Landfill	2.10 (2.76)	2.08 (2.75)	32.71 (2.55)							
Treatment	1.85 (1.95)	2.99 (2.76)	35.76 (2.70)							
On-site										
Landfill	0.89 (1.22)	0.51 (0.68)	26.54 (2.02)	26.43 (2.01)	26.10 (1.96)	31.64 (2.21)	34.42 (2.14)	27.62 (2.03)	34.14 (1.93)	36.93 (1.82)
Treatment	2.78 (3.93)	3.05 (4.29)	25.24 (2.02)	25.09 (2.01)	24.31 (1.91)	29.78 (2.24)	30.98 (2.07)	22.97 (1.82)	30.20 (1.82)	31.49 (1.69)

	Spec 1	Spec 2	Spec 3	Spec 4	Spec 5	Spec 6	Spec 7	Spec 8	Spec 9	Spec 10
Off-site × log volume				−3.00 (−2.54)	−2.94 (−2.44)	−3.42 (−2.73)	−3.88 (−2.61)	−2.80 (−2.36)	−3.63 (−2.23)	−3.93 (−2.11)
Off-site landfill × log volume		−2.98 (−2.51)								
Off-site treatment × log volume		−3.40 (−2.63)								
On-site landfill × log volume			−2.43 (−2.01)	−2.42 (−2.00)	−2.38 (−1.94)	−2.89 (−2.21)	−3.33 (−2.15)	−2.53 (−2.04)	−3.17 (−1.91)	−3.57 (−1.85)
On-site treatment × log volume			−2.06 (−1.82)	−2.05 (−1.80)	−1.95 (−1.68)	−2.45 (−2.05)	−2.90 (−2.02)	−1.88 (−1.65)	−2.72 (−1.72)	−3.02 (−1.66)
Off-site × site variable							0.05 (0.91)	2.65 (0.34)	0.11 (1.49)	0.45 (1.74)
On-site landfill × site variable							0.05 (0.75)	−7.06 (−0.70)	0.02 (0.20)	0.16 (0.56)
On-site treatment × site variable							0.08 (1.75)	3.16 (0.42)	0.08 (1.20)	0.35 (1.43)
Log likelihood:	−79.15	−71.63	−62.88	−63.51	−62.73	−62.11	−61.47	−61.86	−61.86	−61.34

Notes: T-ratios appear in parentheses. Specification 1 estimates the choice of cleanup option expressed as a function of cost and technology dummies. Specifications 2 and 3 include interactions of the explanatory variables with a measure of volume. Specifications 4, 5, and 6 aggregate the off-site technology dummies into a single off-site category and include interactions of the cost variables with measures of site risk. Specifications 7, 8, 9, and 10 expand on Specification 4 by including interactions of the technology dummies with certain site characteristics.

[a] Air route score not included.
[b] Income is measured in thousands of 1989 dollars for the zip code in which the site is located.
[c] Lvol1 = min[log(volume), log(15,000)]; volume is in cubic yards.
[d] Current soil risk = excess lifetime cancer risk, plausible maximum case.

nologies at all waste volumes in the sample. Off-site disposal (there is no difference in the value attached to off-site landfills versus off-site treatment) is the second-most preferred option at sites of 15,000 cubic yards or less.

With the exception of volume, the choice of cleanup option at PCB sites is relatively unaffected by site characteristics, as can be seen in Specifications 4 through 10. In particular, the allegation that EPA has selected less permanent cleanups in minority areas appears false. Interactions of the percentage of the population that is nonwhite with the technology dummies—see Specifications 8 through 10—are insignificant at conventional levels.

The only variables that are marginally significant when interacted with the technology dummies are HRS score and per capita income. EPA was more likely to choose on-site treatment at a site the higher its HRS score.[13] The other interaction term that is marginally significant is the product of per capita income and the off-site dummy. This suggests a preference for off-site treatment in neighborhoods with higher per capita incomes.

The Value of More Permanent Cleanups

Because Table 6 indicates that EPA was willing to pay more for more permanent cleanups, it is interesting to see exactly how large these valuations are. Figure 2 shows the value attached to different cleanup options by size of site, based on Specification 2 of Table 6. At a 10,000-cubic-yard site, EPA would be willing to pay $12.1 million (1987 dollars) to treat waste on-site rather than contain it. For sites with 15,000 or more cubic yards of contaminated waste, however, this figure jumps to $36.5 million.[14] The values attached to off-site treatment (compared to containment) are almost as large—$11.9 million for sites of 10,000 cubic yards and $35.8 million for sites in excess of 15,000 cubic yards.

Off-site disposal of excavated soil was also valued positively by the agency—indeed, the value of transporting waste off-site rather than containing it on-site was $8.25 million at a site of 10,000 cubic yards and $24.8 million at a site containing 25,000 cubic yards of waste. This implies that the agency implicitly valued off-site landfilling of waste more than on-site landfilling (whose coefficient is not significantly different from zero), an interesting result in view of the preference of the SARA amendments for on-site disposal. The more important question that Figure 2 raises, however, is whether the implicit valuations of more permanent cleanups agree with amounts that society would be willing to pay for such cleanups.

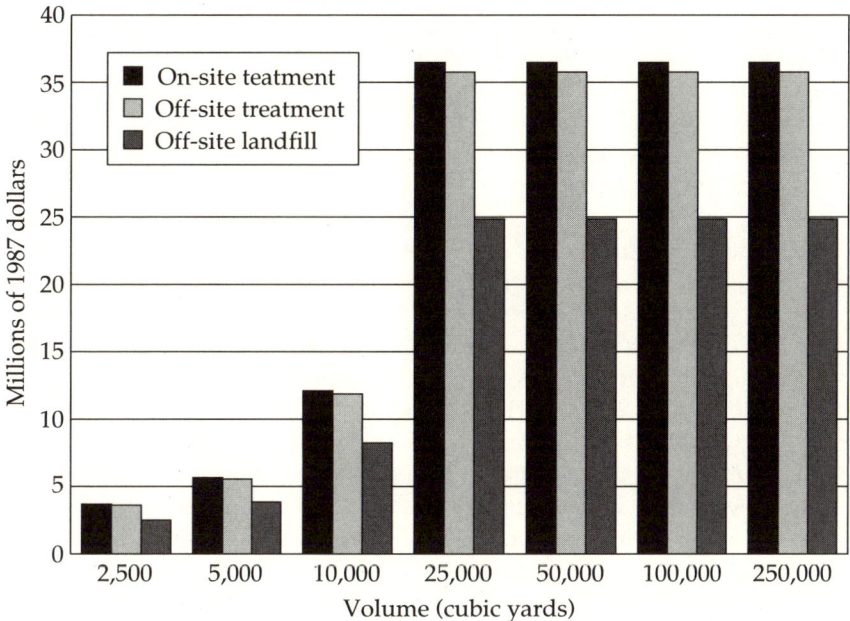

Figure 2. Implicit Valuation of Remedial Options with Respect to Nonexcavation Option (Based on Specification 2, Table 6)

SUMMARY AND CONCLUSIONS

From the perspective of public policy, it is important to ask whether the benefits of Superfund cleanups justify the costs. Because the information presented in site studies (the RI/FSs) is not sufficient to conduct a conventional cost-benefit analysis of cleanup decisions, we attempted to determine whether benefits and costs were correlated with the agency's choice of cleanup goal (target risk level) and choice of cleanup technology.

Our analysis of the choice of target risk level suggests two important findings:
- At both wood-preserving sites and PCB sites, there is no evidence that EPA selected higher target risks at larger sites or at sites with higher costs of excavating and treating contaminated soil. Thus, the agency appears to have obeyed its mandate to protect health without considering costs.
- There is, likewise, no evidence that EPA selected higher target risks in areas with a higher percentage of minority residents or with lower median incomes. Indeed, at wood-preserving sites there is

evidence, after controlling for other variables, that lower target risks were selected at sites with a higher percentage of minority residents. To the extent that the racial composition of the zip code in which the site is located captures racial composition near the site, fears of racial discrimination in choice of cleanup goals are not confirmed by these results.

In addition to these findings, it is interesting to note that, at wood-preserving sites, higher target risks were selected at sites where PRPs were in charge of the RI/FS, thus lending some support to the notion that PRPs can influence the cleanup chosen when they oversee the RI/FS. Higher target risks were also selected at sites with lower surface water HRS scores and at sites in urban areas. The first of these findings may reflect the fact that fewer people are exposed to a hazard the lower the HRS score. Regarding the finding on urban areas, there may be less concern for health at sites in urban areas than at sites in suburban or rural areas if the former are more likely to be in industrial areas.

The second part of our analysis explains the choice of technology used to clean up contaminated soils at the site. The question we initially asked was whether EPA had considered cost in choosing a remedial option, as it is allowed to do by law once a safe target risk level has been chosen, and whether the agency was willing to pay more for more permanent cleanups. There is indeed evidence that EPA considered costs and showed a preference for more permanent remedies at the sites studied. Other things equal, EPA was more likely to select a cleanup option the lower its cost, at all sites examined. EPA also showed a preference for more permanent cleanups. At both sets of sites, EPA was willing to pay (or have others pay) more for on-site incineration of soil than for capping of the soil.

We also investigated two other issues: whether EPA preferred on-site to off-site cleanups and whether the characteristics of the site influenced the remedy selected. As far as the on-site/off-site choice is concerned, EPA was willing to pay more for on-site incineration than for any type of off-site treatment; however, off-site landfilling or incineration of contaminated soil was considered preferable to on-site landfilling of soil at PCB sites. At wood-preserving sites, EPA was indifferent between an on-site landfill and off-site treatment.

The only evidence that site characteristics influenced the remedy selected came at PCB sites. At these sites, off-site treatment of contaminated soil was more likely the higher the per capita income of persons living near the site. This result, however, is only marginally significant in the statistical sense. EPA was also more likely to choose on-site treatment (incineration) at sites with higher HRS scores than at sites with lower HRS scores. In no equation was there any evidence that the racial composition of the zip code influenced the permanence of the remedy selected.

Perhaps the most disquieting of our results is the implicit value attached to permanence in selection of a cleanup option. While EPA showed a preference for more permanent cleanups, as it is supposed to under the SARA amendments, its implicit willingness to pay (or have others pay) for more permanent cleanups was high. At a wood preserving site where capping of contaminated soil would cost $400,000 (1987 dollars), EPA was implicitly willing to spend an additional $11.4 million to excavate and incinerate the soil. The implicit value of incineration over and above the value of capping at PCB sites was $12 million (1987 dollars) at small sites (10,000 cubic yards) and $40 million at large sites (greater than 15,000 cubic yards).

What must be asked is whether the benefits of more permanent cleanups—such as those achieved by incineration of contaminated soil—are worth the amount EPA is willing to pay for them. To answer this question, it will be necessary first to define and then value the benefits of alternative waste disposal technologies. In view of the size of the resources devoted to Superfund cleanups, research into this and related questions deserves the very highest priority.

APPENDIX: A FORMAL MODEL OF CLEANUP DECISIONS

This appendix presents a model of the decision to clean up contaminated soils at Superfund sites. Formally, the cleanup decision is a mixed discrete-continuous choice problem (Hanemann 1984): For each possible technological option, an optimal target concentration is chosen; then the optimal technological choice is made. For each technological option i, lowering the target concentration C_i will raise the number of lives saved from the cleanup, but also will raise the cost. The optimal target concentration for technological alternative i is the C_i that maximizes benefits minus costs, B_i,

$$B_i = L[d(C_i)R^0, P, \mathbf{Z}] + H[t_i, P, \mathbf{Z}] - F[V(C_i, Q)p_i, \mathbf{Z}] + G(v, C_i) + u_i. \quad (1)$$

The first term in Equation 1, L, represents the utility of lives saved by the cleanup. In its most general form, this term is a function of the reduction in risk of death to each person living near the site, as well as the size of the exposed population, P.[15] The reduction in risk of death is the product of baseline risk, R^0, and $d(C_i)$, the percentage reduction in risk of death achieved by concentration C_i. The utility received from saving lives may depend on characteristics of the site, **Z**, which include characteristics of the population living near the site.

The second term in Equation 1, H, measures the benefits of a more permanent cleanup. These benefits are assumed to depend on a measure of permanence, t_i, and possibly on the number of persons, P, living near the site, and **Z**, but are assumed to be independent of target concentration. Our justification for this assumption is that the benefits of more permanent cleanups are not simply an increase in the period over which risk reductions accrue. Increased permanence appears to be valued more for passive-use reasons—the desire to leave a cleaner environment to one's children or to preserve the environment for its own sake.

The third term in Equation 1, F, measures the cost of cleanup, which is the product of the volume, V, of soil addressed and the unit cost, p_i of treatment and/or containment. The volume of soil addressed is a decreasing function of target concentration but an increasing function of the size of the site, Q. Unit treatment costs p_i vary according to the technology used. The last two terms in the net benefit equation represent, respectively, the benefits of a lower target concentration and of a more permanent cleanup, that are not observed by the researcher.

Conditional on i, the optimal C_i is given by

$$C_i = C_i(R^0, P, Q, p_i, \mathbf{Z}, v). \tag{2}$$

Equation 2 can be estimated by least squares to explain variation in target concentrations. Since, in reality, there is a vector of C_i's, one for each pollutant, it is more convenient to treat the target risk level, $R_T = [1-d(C_i)]R^0$, as the dependent variable. This yields the target risk equation

$$R_T = R_T(R^0, P, Q, p_i, \mathbf{Z}, v). \tag{3}$$

In theory, one would estimate Equation 3 using all of the observations on each technological alternative i for each site j.

To analyze the choice of technology, the optimal concentration must be substituted into Equation 1 to yield

$$B_i = U_i + e_i, \tag{4}$$

where U_i is the deterministic portion of net benefits and e_i is the random portion. The probability that cleanup option i is selected is given by

$$P(U_i + e_i > U_k + e_k, \text{ all } k), \tag{5}$$

where k indexes cleanup options other than the one selected.

While Equation 5 appears to be a standard discrete choice model, the errors, e_i and e_k will be correlated, in general, by virtue of the common component, v. The multinomial logit model, however, may serve as a good approximation to Equation 5 even if it is not, strictly speaking, applicable (Cropper and others 1993).

Specification of the Discrete Choice Model

As noted above, the choice of cleanup option is made by comparing the net benefits of each option, as indicated in Equation 1. Because the target concentration (or risk level) associated with each option is not observed (it is reported only for the cleanup option chosen), we must omit the first term, L, on the right-hand side of Equation 1 from that equation. The permanence variable t_i is represented by a vector $\mathbf{T}_i$ of dummies corresponding to the categories described earlier (in the section on the classification scheme for cleanup options).[16] Since the value attached to the benefits and the disutility attached to the cost of cleanup may vary according to site-specific characteristics $\mathbf{Z}$, we include terms in the net benefit function that interact site characteristics with cost and with the technology dummies.[17] The resulting discrete choice model is estimated as a multinomial logit model.

ACKNOWLEDGMENTS

This research was funded by the U.S. Environmental Protection Agency through Cooperative Agreement CR-818454-01-0. We would like to thank Kate Probst for her comments and encouragement and for providing us with data. We would also like to thank Bill Evans for his help with econometric matters, and Wally Oates and Terry Davies for helpful discussions on the subject.

ENDNOTES

[1] EPA estimates range from $52 billion for 2,100 sites to $250 billion for 10,000 sites, whereas a study by the Chemical Manufacturers Association (1988) arrives at a range of $54 billion to $133 billion for a 2,000-site NPL. The U.S. Congress' Office of Technology Assessment (U.S. Congress 1989) estimates the cost at $500 billion, assuming the NPL would ultimately consist of 10,000 sites.

[2] Cleanup decisions must also be made with respect to contaminated groundwater; however, the remedy chosen for contaminated groundwater—to pump the water and treat it—varies little from one site to another. We therefore focus on the cleanup decision for contaminated soils.

[3]The thirty-two wood-preserving sites include all wood-preserving sites for which RODs had been signed as of the end of fiscal year 1991. The seventy-eight sites with PCB contamination were sites with PCB contamination in excess of 10 parts per million for which RODs had been signed as of the end of fiscal year 1991.

[4]PCBs are a group of toxic chemicals that, prior to being banned in 1979, were used in electrical transformers, hydraulic fluids, adhesives, and caulking compounds. They are extremely persistent in the environment because they are stable, nonreactive, and highly heat resistant.

[5]The model in the appendix to this chapter implies that raising the baseline risk raises the target risk chosen. If there is diminishing marginal utility from saving lives, the benefits of lowering the target risk level are reduced with any raising of the baseline risk.

[6]The maximally exposed individual may be a child who ingests contaminated soils, a person working at a still-active site, or a resident living within the boundaries of the site.

[7]In some instances, future risk is not specified in a ROD. We handle this problem by setting future risk equal to zero and defining a dummy variable, RM. ($RM = 1$ if future risk is missing.) The missing data indicator also enters as an independent variable. The coefficient of risk therefore represents the effect of risk given that it is known.

[8]A better measure of the size of a site would be volume of contaminated waste in excess of some small concentration; however, this is unavailable. The volume of waste excavated is inappropriate as an exogenous measure of size since it is determined by the target concentration.

[9]In situ treatment includes both flushing the soil, which removes contaminants, and bioremediation, which uses bacteria to neutralize toxic substances.

[10]Interacting volume and cost produces insignificant results at wood-preserving sites.

[11]This is clearly true by inspection in columns 1 and 2 of Table 6. In columns 3 through 10 of the table, it is also true if one evaluates the coefficients of the technology dummies at different volumes of waste.

[12]The excluded category in Table 6 is that of nonexcavation cleanups, which include both containment of waste and in situ treatment. The two categories were combined because in situ treatment was rarely considered at PCB sites.

[13]In Table 6, a modified version of the HRS score is used that combines the surface and groundwater components of the score but eliminates the air score. This is done because it is impossible to distinguish cases in which the air score is zero from cases in which the air score has not been computed (a frequent occurrence if the surface and groundwater scores are sufficient to put a site on the NPL).

[14]Recall that the interaction of cost with log(volume 1) implies that the effect of volume stops at volumes of 15,000 cubic yards. That is, the disutility attached to cost at sites of 15,001 cubic yards is the same as the disutility at sites of 50,000 cubic yards.

[15] If EPA cared only about the number of lives saved, benefits would depend on the product of the reduction in individual risk times the size of the exposed population. It has been suggested, however, that the agency cares about the reduction in individual risk independently of the reduction in lives saved (Travis and others 1987). That is, reducing risk of death by one in a million for each of one million people is not equivalent to reducing risk of death by one in a thousand for each of one thousand people.

[16] In the estimating equation, at most five of the categories can be used, since a constant term is included in the equation.

[17] For example, if EPA has, as alleged by Lavelle and Coyle (1992), a preference for less permanent cleanups in areas with a significant minority population, then the coefficients of the technology dummies will be functions of "Percent nonwhite population." Likewise, site characteristics (such as "Median income" and "Baseline risk") may alter the disutility attached to cost.

REFERENCES

Chemical Manufacturers Association. 1988. *Impact Analysis of RCRA Corrective Action and CERCLA Remediation Programs*. Washington D.C.: Chemical Manufacturers Association.

Cropper, Maureen L., Leland B. Deck, Nalin Kishor, and Kenneth E. McConnell. 1993. Valuing Product Attributes Using Single Market Data: A Comparison of Hedonic and Discrete Choice Approaches. *Review of Economics and Statistics* 75: 225–232.

Hanemann, W. Michael. 1984. Discrete/Continuous Models of Consumer Demand. *Econometrica* 52: 541–561.

Lavelle, Marianne, and Marcia Coyle. 1992. Unequal Protection: The Racial Divide in Environmental Law. *National Law Journal* 15 (September 21): 1–12.

Travis, Curtis, Samantha Richter, Edmund Crouch, Richard Wilson, and Ernest Klema. 1987. Cancer Risk Management: A Review of 132 Federal Regulatory Decisions. *Environmental Science and Technology* 21: 415–420.

United Church of Christ. 1987. Commission for Racial Justice. *Toxic Wastes and Race in the United States, A National Report on the Racial and Socio-Economic Characteristics of Communities with Hazardous Waste Sites*. New York: United Church of Christ.

U.S. Congress. 1989. *Coming Clean: Superfund's Problems Can Be Solved*. OTA-ITE-433. Washington D.C.: Office of Technology Assessment (OTA).

U.S. EPA (Environmental Protection Agency). 1990. *Progress Towards Implementing Superfund, Fiscal Year 1990: Report to Congress*. EPA/540/8-91/004. Washington, D.C.: U.S. EPA, Office of Emergency and Remedial Response.

PART III:
The Liability Regime

5

Evaluating the Effects of Alternative Superfund Liability Rules

Lewis A. Kornhauser and Richard L. Revesz

One of the central issues in the Superfund reauthorization debate concerns the fate of joint-and-several liability. The 1980 Superfund statute did not specify whether potentially responsible parties (PRPs) would face joint-and-several liability rather than nonjoint (several only) liability. In the face of this legislative silence, the courts followed the approach of the Restatement (Second) of Torts and have imposed joint-and-several liability for indivisible harm—harm for which there is not a reasonable basis of apportionment (Kornhauser and Revesz 1989, 851–53). The Superfund Amendments and Reauthorization Act of 1986 (SARA) implicitly ratified the imposition of joint-and-several liability, at least in some circumstances, by establishing a right to contribution; such a right would not arise in the absence of joint-and-several liability.

In the reauthorization debate, two groups recently have urged various curtailments in the use of joint-and-several liability. The National Commission on Superfund, comprised of twenty-six leaders from the various Superfund constituencies, did so in its final consensus report (National Commission on Superfund 1994). The Clinton administration likewise urged such curtailments in its bill submitted to the 103rd Congress in February 1994 (H.R. 3800).

The commission recommended a scheme under which PRPs would generally be responsible only for their own share of the costs and would typically not have to bear responsibility for any orphan shares of insolvent or unidentifiable PRPs, if they accepted a determination of their share performed by a neutral third party. Orphan shares would be funded by an increased tax and would be charged to the PRPs participating in the allocation process only if the proceeds of the tax were insufficient. Joint-and-several liability would be retained for parties that did not comply with the requirements of the allocation process (National Commission on Superfund 1994, 14–21).

Under the Clinton administration's Superfund bill, PRPs that accept a neutral determination of their share of the liability would not be responsible for amounts attributable to insolvent PRPs, but they would have to pay their share of amounts attributable to unidentifiable PRPs. Even under the current Superfund statute, the courts have disagreed somewhat about what it means for a harm to be indivisible, thereby triggering joint-and-several liability.

This chapter seeks to inform the reauthorization debate by comparing the effects of joint-and-several liability to those of nonjoint liability on the basis of three criteria: deterrence, settlement-inducing properties, and fairness. Our major conclusion is that, with respect to each of these criteria, neither rule dominates the other.

Our discussion in this chapter focuses on a simple situation: a single plaintiff, whether the U.S. Environmental Protection Agency (EPA) or a state, seeks to recover cleanup costs from two defendants; we therefore do not focus on many of the real-world complications posed by Superfund. We proceed in this fashion because it is important to understand the basic questions raised by joint-and-several liability before dealing with additional complications. The major insights of this chapter are applicable to more complex situations as well.

DETERRENCE

We consider in this first part the relative effects of joint-and-several liability and nonjoint liability in transmitting to generators of hazardous wastes appropriate incentives for waste reduction and care. The first section of our discussion on this topic covers the role of liability rules in transmitting incentives for desirable conduct concerning wastes generated both before and after the passage of Superfund. The second section presents the model that guides our inquiry. The third section analyzes the situation in which the generators have sufficient solvency to satisfy the judgments entered against them; it shows that joint-and-several liability and nonjoint liability have identical effects but that both, when coupled with strict liability, produce underdeterrence. The fourth and final section studies how the situation is different when one of the defendants has limited solvency; it shows that, under these circumstances, it is not possible to draw any general conclusion about whether joint-and-several liability is preferable to nonjoint liability. (A more technical treatment of the issues discussed in these last two sections can be found in Kornhauser and Revesz 1989, 1990.)

The Role of Liability Rules

The relative desirability of competing regimes' imposing liability on generators of hazardous wastes should be evaluated by reference to the incentives that the regimes transmit for desirable conduct. Generators make several relevant decisions that can be influenced by liability rules; the following list is by no means exhaustive.

First, a generator must determine the volume of hazardous wastes that it will produce. Because the wastes are generated as a byproduct of profitable economic activity, this decision is affected by the costs of different production processes as well as the expected liability associated with the generation of hazardous wastes. So, for example, a generator must trade off the costs of more expensive production processes that would yield a smaller amount of wastes per unit of useful output against the higher costs, including the expected liability, of disposing a larger amount of wastes.

Second, a generator must determine whether to recycle, to treat (that is, render nonhazardous), or to dispose of the wastes. Each of these processes has certain immediate costs and gives rise to different levels of expected liability.

Third, a generator must choose a level of care for the handling of the wastes in the predisposal phase. More care raises the cost of disposal but decreases the generator's expected liability.

Fourth, for wastes that it chooses to dispose, a generator must choose a disposal site. Different sites might charge different fees, present different risks that at some point hazardous wastes will be released into the environment, and give rise to different cleanup costs in the event of a release.

Fifth, both in the case of on-site and off-site disposal, a generator must decide on the effort that it will expend in monitoring the disposal site to detect releases of the wastes into the environment. In fact, such monitoring has led to the cleanup of a number of sites that are not on the National Priorities List (NPL). Once a release occurs, cleanup costs might rise quickly as a function of the time that the problem is left unattended. Thus, by monitoring a site and perhaps undertaking a cleanup before the site has come to the attention of EPA, a generator can reduce its liability. Here, a trade-off exists between the monitoring costs and the expected liability.

Sixth, once a release is detected—whether by the generator, by another PRP, or by EPA or a state—a cleanup has to be undertaken. A liability scheme can provide incentives for the generator to act to ensure that the cleanup is performed in a cost-effective manner.

In analyzing the effects of liability rules, the Superfund program should be separated into two distinct components: a prospective component, for wastes generated and disposed after the passage of the statute in 1980; and a retrospective component, for wastes generated and disposed prior to the passage of the statute.

Obviously, with respect to the retrospective component, a liability scheme cannot create incentives for actions that have already been completed. They can, however, provide incentives for monitoring the site and for ensuring that a cleanup is performed in a cost-effective manner. Opponents of retroactive Superfund liability typically overlook these incentives.

With respect to monitoring, it would be desirable if PRPs undertook cleanups at non-NPL sites before the site came to the attention of EPA or the state (Probst and Portney 1992, 15). Not only would EPA then be able to better deploy its scarce enforcement resources, but environmental problems could be addressed earlier, before the cleanup costs escalated. This problem of cost escalation is particularly acute under the Superfund program, where groundwater remediation comprises a large part of cleanup costs. Such remediation is less likely to be necessary, or its extent reduced, if the cleanup is undertaken soon after the problem is discovered.

With respect to cleanup decisions, there is strong evidence suggesting that, for a given cleanup level, the cleanup costs are lower when the actions are undertaken by PRPs rather than by EPA; current estimates of this differential run at around 20%. Moreover, EPA is constrained in the number of sites at which it can supervise cleanups. It has no alternative but to do so at sites that have no PRPs with the resources or expertise to perform cleanups. For other sites, it relies heavily on PRPs, which now perform about 70% of the cleanups. The vast majority of the PRPs currently undertaking cleanups generated hazardous wastes before the passage of Superfund. If these parties did not face liability, the cleanups would have to be performed by EPA. Thus, in the absence of retroactive liability, not only would cleanup costs be higher, but our ability to address the problem of contaminated sites would be impaired.

The Model

In our analysis we model a situation in which two manufacturers, Row and Column, dispose of hazardous wastes at a single landfill. These actors benefit from the dumping because the wastes are the byproduct of profitable economic activity. At some time in the future, these wastes may leak into the environment and cause serious damage, including, perhaps, the contamination of groundwater supplies. For ease of exposition, we think of this damage as the cost of cleaning up the landfill and the surrounding area affected by the release. We assume initially that, as the volume of wastes at the landfill grows, this cost increases more rapidly than linearly. We then show that this assumption of convex cleanup costs is not necessary for our central result, which holds even if the cleanup costs are linear or concave.

The expected damage from a release does not fall directly on the generators unless a legal provision shifts the liability to them. Instead, it falls on those who would either undertake a cleanup or, alternatively, suffer the consequences if the problem were left unattended. In our discussion, we assume that EPA initially bears the cleanup costs and then seeks reimbursement to the extent allowed by the liability regime.

The efficient amount of wastes is that which maximizes the social objective function: the sum of the benefits derived by the manufacturers minus the expected damages. An economically rational generator, however, does not make its decision based on the social objective function. Instead, it seeks to maximize its private objective function: the benefit that it derives from the activity that leads to the production of the wastes minus whatever share of the damage allocated to it. This share depends on the liability regime that applies in the case of a release of hazardous wastes. For expositional convenience, we assume that the only way in which an actor can affect its expected liability is through the quantity of wastes that it chooses to generate. The results developed below, however, extend to the more complex situation discussed in the previous section on the role of liability rules.

Superfund has adopted a regime of strict liability as opposed to negligence. For a rule of strict liability, we compare joint-and-several liability with nonjoint liability, first when Row and Column both have sufficient solvency to pay their share of the damage (the full solvency case) and then when Row's solvency is limited. Under a rule of joint-and-several liability, EPA could recover its entire judgment from either one of the defendants. That defendant, having paid the full cleanup cost, would then have to resort to a *contribution action* in order to recover the other's share of the cost. We assume, instead, that the contribution action is joined with EPA's claim against the two generators, and that EPA recovers from each in proportion to its share of the liability.

The argument can best be developed by reference to a simple example. Let x and y be the amount of wastes generated by Row and Column, respectively. Let $(100 + 20x)$ and $(100 + 20y)$ be the benefits that Row and Column, respectively, obtain from engaging in the economic activity that produces wastes as a byproduct. (If either Row or Column exits the market, it does not receive the fixed component of $100, but the remaining actor receives that additional amount; one can thus think of the market as guaranteeing a $200 profit, which is either split by two firms or captured by a single firm.) Let $(x + y)^2$ be the damage from the disposal of these hazardous wastes. Net social benefits are maximized where $(x + y) = 10$; thus, one efficient outcome is for Row and Column to generate 5 units of wastes each.

We can now illustrate the difference between joint-and-several liability and nonjoint liability under both strict liability and negligence regimes. Under strict liability, as long as the defendants are sufficiently solvent, joint-and-several liability transmits the same incentives as nonjoint liability. Under either rule, each defendant would pay its share of the total damage. So, for example, if Row generated 4 units of wastes and Column generated 6 units, Row would pay 40% of the total damage of $100, or $40, and Column would pay 60% of this amount, or $60. Joint-and-several liability and nonjoint liability produce the same consequences because there is no scenario under which either defendant would be called upon to pay costs attributable to the other.

In contrast, the choice between joint-and-several liability and nonjoint liability matters under strict liability when at least one of the defendants has limited solvency. If Row were wholly insolvent, Column's liability would be unaffected under nonjoint liability; it would still pay $60. Under joint-and-several liability, Column would instead pay the full damage of $100.

The choice between joint-and-several liability and nonjoint liability matters under negligence if one defendant is negligent and the other is not, or if both are negligent and only one is solvent. Assume, in the preceding example, that either Row or Column will be deemed negligent if it generates more than 5 units of wastes; we denote this level of waste generation as x^*. Row, which generated only 4 units, is nonnegligent and therefore not liable. Under joint-and-several liability, Column is then responsible for the full damage of $100: $60 attributable to its own wastes and $40 attributable to the wastes of its nonnegligent cogenerator. In contrast, under nonjoint liability, Column is responsible only for $60 and the plaintiff would not recover the damage attributable to the nonnegligent generator.[1]

Full Solvency

Having established that, for strict liability, joint-and-several liability has the same effect as nonjoint liability if the defendants are sufficiently solvent, we show that strict liability—regardless of whether it is coupled with joint-and-several liability or nonjoint liability—fails to transmit desirable incentives: it leads to the overproduction of wastes. Recall that social welfare is maximized when Row and Column each generates 5 units of wastes. Assume that Row has tentatively decided to generate 5 units and that Column, without consulting Row, is trying to figure out how much to generate. If it also generated 5 units, it would accrue benefits of $200 and face a liability of one-half the total damage of $100, or $50; its net benefits would therefore be $150. What would happen, however, if Column generated 6 units rather than 5? Its benefits would rise from $200 to $220, its share of the damages would rise from one-half to six-elevenths (6/11), and the total damages would rise from $100 to $121. In sum, Column's net benefits would be $154 rather than $150.[2]

In turn, Column will have imposed a cost on Row. Row's share of the damages will fall from one-half to five-elevenths (5/11), but, as a result in the increase in damages, it will have to pay $55, rather than $50. Its net benefits will fall from $150 to $145. Thus, by deciding to generate 6 units of wastes rather than 5, Column captures an additional $4 in net benefits, but imposes costs of $5 on Row. As a result, Column's action decreases the aggregate level of net social benefits.

Of course, Row can play the same game too. The symmetric Nash equilibrium of this game (the point at which neither Row nor Column has an incentive unilaterally to change its strategy) occurs when each of these actors generates 6.67 units of wastes. We will denote this level as $x(\infty)$, which is larger than x^* (the socially optimal level of wastes generated). The net benefits of each of these parties is then $144.42, rather than the $150 that each would have accrued if they both had acted in the socially optimal fashion.

Thus, strict liability (coupled with either joint-and-several liability or nonjoint liability) underdeters: it leads to the production of an excessive amount of wastes. (We stress, however, that this underdeterrence is a product of the convexity of cleanup costs.) This problem could be averted if Row and Column acted cooperatively. Then, they could agree to each generate 5 units of wastes instead of 6.67 units, and each would obtain net benefits of $150 rather than $144.42. In the case of generators of hazardous wastes, such cooperation is probably unrealistic because large numbers of actors are involved and they make their decisions at different times.

The inefficiency that we illustrated by reference to a specific example is a general feature of strict liability in the context of joint tort-feasors. This inefficiency stems from the fact that an actor that contemplates generating more than the socially optimal level of wastes, given that the other actor is generating the socially optimal level, does not bear the full increase in damages that it imposes on society. Instead, the other generator bears part of this damage. The resultant externality produces the socially suboptimal results.

A rule of negligence avoids this inefficiency when coupled with joint-and-several liability, as long as the standard of care is set at the socially optimal level. Assume, again, that Row has tentatively decided to generate 5 units of wastes and that Column, without consulting Row, is trying to figure out how much to generate; the standard of care is 5 units for each of these actors. If Column also generated 5 units, it would accrue benefits of $200 and face no liability at all, as it would meet its standard of care; the total damages of $100 would go uncompensated. What would happen, however, if Column generated 6 units rather than 5? Its benefits would rise from $200 to $220, and the total damages would rise from $100 to $121. Under negligence, however, Column would be responsible for the full damages: both the amount attributable to its negligent conduct and that attributable to the nonnegligent conduct of its cogenerator. Thus, generating 6 units rather than 5 reduces Column's net benefits from $150 to $99.

The reason that negligence coupled with joint-and-several liability transmits the correct incentives is quite straightforward. When one actor is not negligent, joint-and-several liability assigns to a negligent actor two components of damages: the additional damage caused by its negligence and the damage that would have resulted if both actors acted nonnegligently, which otherwise would have been borne by the plaintiff. The nonnegligent actor (Row in our case) neither gains nor loses from the decisions of its cogenerator (Column) to become negligent, because it bears none of the loss and receives none of the resulting benefits. The negligent actor, on the other hand, would capture the full additional benefits of generating more than the standard of care, but would pay for more than the additional damage. Since the standards of care are set so that social welfare is maximized when all the actors meet the standard of care, it follows that the additional benefit captured by Column is smaller than even the additional damage caused by its negligence. As a result, both actors choose to be nonnegligent and, therefore, generate at the level that maximizes social welfare. Negligence, it should be noted, does not necessarily provide the right incentives when it is coupled with nonjoint liability (see Kornhauser and Revesz 1989).

This argument about the more desirable properties of negligence in the context of our model of fully solvent actors is presented solely to underscore more clearly the problems with strict liability. Plausible reasons for not adopting a rule of negligence include lack of confidence in the decisionmaker's ability to set the standard of care at the socially optimal level and the transaction costs necessary to assess the care taken by the various actors.

Limited Solvency

We assume here that each actor has a fixed solvency, which is available to pay the actor's share of the social loss. We assume that the benefits that the actor derives from engaging in the activity leading to the production of hazardous wastes are not included in this solvency. We offer two interpretations for this assumption. First, one might think of an actor's solvency as a bond that it must post in order to engage in the activity. Recourse against the actor is then restricted to the size of the bond. Alternatively, one might interpret the model as containing an implicit time structure. Benefits accrue to the actor in the present while the social loss occurs (and responsibility for it is apportioned) in the future. If current profits are distributed in the present (when they accrue), then the actor's solvency has the traditional interpretation of the difference between its assets and liabilities.

To analyze the problem of limited solvency, we modify in one respect the model presented previously: we assume that that the actors derive benefits from the level of hazardous wastes generated only up to a technological limit, which we call x^H. Beyond x^H, no further benefits accrue from additional waste generation. Let x^H be equal to 9 units. We look only at the particular case in which Row's solvency is zero and examine the properties of strict liability for different values of Column's solvency. We restrict our inquiry in this fashion because it is sufficient to generate our central conclusion: that it is not possible to make general comparisons between joint-and-several liability and nonjoint liability.

Joint-and-Several Liability Equilibria

Recalling that Row's solvency is zero, consider first a case in which Column is fully solvent. Under joint-and-several liability, Column will be responsible for the whole liability. If Row were generating x^*, Column would generate x^* as well, as it would incur the full social cost of departing from x^*. (As in the prior section on full solvency, we defined x^* as the socially optimal amount of waste generation; in our example, x^* is equal to 5 units.) But because Row is generating at the technological limit of x^H, which is more than x^*, Column's best response is to generate less than x^*—an amount that we shall call a. (The first symbol is Row's level of waste generation and the second is Column's.) Therefore, the resulting equilibrium is (x^H, a). In our example, a is the value of y that maximizes $100 + 20y - (y+9)^2$ and is equal to 1 unit. Column's net benefits are therefore $20.

If Column is not fully solvent, there are two possible equilibria: (x^H, a) or (x^H, x^H). Column will generate a (the equilibrium will be at (x^H, a)) if Column has sufficient solvency to pay for the liability attributable to it if it generated x^H, which is the full damage caused by both parties' generating a total of $2x^H$. In contrast, the equilibrium is at (x^H, x^H) if Column does not have sufficient solvency to pay for the liability attributable to it if it generated a, which is the full damage caused when both parties generate a total of $(x^H + a)$.

But the equilibrium can be at (x^H, x^H) even if Column has sufficient solvency to pay the liability that results from the equilibrium (x^H, a). Column's additional liability caused by generating more than a is greater than its corresponding additional benefit, up to the point at which it becomes insolvent. Beyond its point of insolvency, however, it continues to accrue benefits, but its liability does not increase.

We define Column's *critical solvency* under a strict liability coupled with joint-and-several liability (given that Row is insolvent) as $s_{C_{sj}}$; this is the lowest solvency for which Column would choose a (1 unit) rather than x^H (9 units) in light of Row's choice of x^H. In our example, if Column generates 9 units of wastes, its benefits are $280. Provided that its solvency is less than $260, its net benefits from generating 9 units are greater than its net benefits from generating 1 unit; $s_{C_{sj}}$ is therefore equal to $260 ($280 minus $20).

To understand more fully the impact on Column of Row's insolvency, we need to look at how Column's solvency would have affected its decision if Row had been fully solvent. Recall that the full solvency equilibrium is $(x(\infty), x(\infty))$—in our example, Row and Column each generates 6.67 units of wastes; under this equilibrium, each actor derives net benefits of $144.42.

If Row is generating 6.67 units of wastes, when would Column prefer to generate 9 rather than 6.67 units? Column would not do this if it were fully solvent, since it would derive $280 in benefits, but would be responsible for a share of 9/15.67 of a total liability of $(15.67)²—it would have to pay $141.03 and its net benefits would therefore be only $138.97 rather than the $144.42 that it would get by generating 6.67 units. On the other hand, if Column's solvency were less than $135.58, Column's net benefits would be greater when it generated 9 rather than 6.67 units; Column would then obtain benefits of $280 and pay only its solvency, thereby obtaining more than $144.42. We define Column's critical solvency under a strict liability coupled with joint-and-several liability, given that Row is fully solvent, as s_s. In this example, s_s is thus equal to $135.58 ($280 minus $144.42).

Not surprisingly, the critical solvency $s_{C_{sj}}$ is greater than s_s. Indeed, for a given level of waste generation, Column must bear a higher liability when Row has zero solvency and is therefore generating x^H than when Row is solvent and is generating $x(\infty)$. First, for any level of waste generation by Column, the total liability is greater where Row is generating x^H than where it is generating $x(\infty)$. Second, when Row has zero solvency, Column is responsible not only for its own share of the liability, but for Row's share as well.

Thus, Column exhausts a given solvency at a lower level of waste generation when Row is insolvent, leaving Column with a greater range in which it can continue to expand its output with no corresponding increase in liability. As a result, there are instances in which Column would not become insolvent and would not expand its output to x^H if Row were solvent, but where Column becomes insolvent and expands its output to x^H if Row is insolvent. Under this "domino effect," the insolvency of Row is the but-for cause of Column's insolvency.

In our example, if Column's solvency is less than $135.58, Column will become insolvent regardless of Row's solvency, and if Column's solvency is greater than $260, Column will remain solvent regardless of Row's solvency. If, however, Column's solvency is between $135.58 and $260 (that is, between s_s and $s_{C_{sj}}$), it will remain solvent if Row also is solvent, but become insolvent if Row is insolvent.

Nonjoint Liability Equilibria

Under nonjoint liability, Column would only be responsible for its apportioned share of the liability, and the plaintiff would not be compensated for the share attributable to Row if Row is insolvent. Even under nonjoint liability, however, Row's insolvency affects Column's incentives. Where Row is insolvent and generates at the technological limit of x^H, Column faces higher costs than where Row is solvent and generates $x(\infty)$. This result follows from the assumption that cleanup costs, as a function of the volume of wastes, increase more rapidly than linearly. Consequently, if Column were fully solvent, it would, in response to Row's insolvency, restrict the amount of wastes that it generates to b, which is less than $x(\infty)$. In our example, Column is responsible for a share of $y/(y+9)$ of the total damages $(y+9)^2$. Then b, the value of y that maximizes $[100 + 20y - [y/(y+9)](y+9)^2]$, is equal to 5.5 units. At this level of waste generation, Column accrues benefits of $210 and must pay $79.75; its net benefits are therefore $130.25

By analogy to the analysis performed for joint-and-several liability, it follows that under nonjoint liability, there is either an equilibrium at (x^H, b), where Column remains solvent, or (x^H, x^H), where Column is insolvent. Our example illustrates a general proposition: because for a given level of waste generation, Column bears less liability under nonjoint liability than under joint-and-several liability, it follows that b is always greater than a.

We define Column's critical solvency under strict liability, this time coupled with nonjoint liability, given that Row is insolvent, as $s_{C_{sn}}$. In our example, if Column generates 9 units, its benefits are $280. Provided that Column's solvency is less than $149.75 ($280 minus $130.25), its net benefits from generating 9 units are greater than its net benefits from generating 5.5 units; $s_{C_{sn}}$ is therefore equal to $149.75.

This critical solvency is greater than s_s, Column's critical solvency under a strict liability coupled with joint-and-several liability, given that Row is fully solvent. The "domino effect" discussed in the context of joint-and-several liability is therefore also present here; the insolvency of Row can be the but-for cause of Column's insolvency even under nonjoint liability.

The critical solvency $s_{C_{sn}}$ is smaller than $s_{C_{sj}}$, Column's critical solvency (given that Row is insolvent) under strict liability coupled with joint-and-several liability. The reason is that if Row is insolvent and Column is solvent, Column faces a smaller liability under nonjoint liability, and the prospect of remaining solvent is correspondingly more attractive. Thus, there is a range of solvencies—the range between $s_{C_{sn}}$ and $s_{C_{sj}}$, between $149.75 and $260 in our example—in which Column would choose to remain solvent under nonjoint liability but would become insolvent under joint-and-several liability.

Comparison of Equilibria under Joint-and-Several Liability and Nonjoint Liability

Table 1 summarizes the equilibria generated by joint-and-several liability and nonjoint liability. Recall that Row's solvency is zero. This table defines three regions for different ranges of Column's solvency. It reveals that in region A, the two rules perform identically, yielding an equilibrium at (x^H, x^H). In region B, joint-and-several liability produces an equilibrium at (x^H, x^H) whereas nonjoint liability does so at (x^H, b). Finally, in region C, the equilibrium under joint-and-several liability is at (x^H, a), whereas under nonjoint liability it is at (x^H, b).

We use three measures to compare the performance of the two rules. We determine, first, which rule leads to higher social welfare; second, which rule results in less unfunded liability; and, third, which rule leads to the generation of less waste. Obviously, in region A, both rules produce identical results.

In region C, we ascertain whether an equilibrium at (x^H, a) under joint-and-several liability is preferable to an equilibrium at (x^H, b) under

Table 1. Equilibria Generated by Joint-and-Several and Nonjoint Liabilities

		Equilibria	
Region	Column's solvency	Joint-and-several liability	Nonjoint liability
A	$0 - s_{C_{sn}}$	(x^H, x^H)	(x^H, x^H)
B	$s_{C_{sn}} - s_{C_{sj}}$	(x^H, x^H)	(x^H, b)
C	$s_{C_{sj}} - \infty$	(x^H, a)	(x^H, b)

nonjoint liability. From a social welfare perspective, (x^H, a) is preferable. Where one actor is generating x^H, joint-and-several liability makes the other actor see the full social cost of its actions, whereas nonjoint liability does not. Thus, a is the optimal response by Column to Row's choice of x^H.

Joint-and-several liability also results in less unfunded liability. When the equilibrium is at (x^H, a), Column is solvent and pays the full liability, leaving no unfunded liability. At (x^H, b), Column is also solvent, but under nonjoint liability, it pays only its share of the liability, leaving Row's share unfunded.

Finally, the (x^H, a) equilibrium results in an amount $(x^H + a)$ of wastes, whereas the (x^H, x^H) equilibrium results in $2x^H$ of wastes. Because a is smaller than x^H, joint-and-several liability is preferable to nonjoint liability. Thus, in region C, joint-and-several liability is preferable under all three criteria.

The comparison between the equilibria in region B follows by analogy. From a social welfare perspective, the equilibrium under nonjoint liability, (x^H, b), is preferable to the equilibrium under joint-and-several liability, (x^H, x^H), because, as indicated, given that Row has chosen x^H, the socially optimal response by Column is a; of course, b is closer to a than is x^H. In terms of amount of waste generated, the (x^H, b) equilibrium is also preferable, as $(x^H + b)$ is smaller than $2x^H$.

With respect to unfunded liability, if Column's solvency is only $s_{C_{sn}}$, which is the lower bound of the range defined in region B, the equilibrium under nonjoint liability, (x^H, b), will result in less unfunded liability. Indeed, as Column expands its production of wastes from b to x^H, the total liability increases, but Column's contribution to that liability does not, as its full solvency is consumed at the (x^H, b) equilibrium. Where Column has a solvency higher than $s_{C_{sn}}$, however, it is not possible to make any general comparison between the rules: the equilibrium under nonjoint liability, (x^H, b), will result in less unfunded liability for certain benefit-and-loss functions, but in more unfunded liability for other functions.

In summary, neither rule dominates the other. In region C, joint-and-several liability is always preferable under all three criteria. In region B, nonjoint liability is always preferable under two criteria—social welfare and waste generated—but is only sometimes preferable under the criterion of unfunded liability. Table 2 summarizes these results.

The result—that when one of the defendants has limited solvency, neither rule dominates the other—does not depend upon the assumption that cleanup costs are convex. If they were concave (or linear), it would nonetheless be the case (from a social welfare perspective) that

Table 2. Comparison of Joint-and-Several Liability with Nonjoint Liability

Region	Column's solvency	Preferred rule
A	$0 - s_{C_{sn}}$	Indifference
B	$s_{C_{sn}} - s_{C_{sj}}$	Nonjoint[a]
C	$s_{C_{sj}} - \infty$	Joint-and-several

[a]For a certain portion of the range, and for certain benefit and loss functions, joint-and-several liability is preferable from the standpoint of unfunded liability.

joint-and-several liability would be preferable, because it would show Column the full cost of its actions, as long as it did not cause Column to become insolvent as well. (Note, however, that if cleanup costs are concave, the benefits of waste generation must be even more concave or else the socially desirable amount of waste would be infinite.) In contrast, if the existence of joint-and-several liability is the but-for cause of Column's insolvency, nonjoint liability would be preferable.

SETTLEMENT-INDUCING PROPERTIES

To evaluate the effects of joint-and-several liability on settlements, we need to specify with care the relevant elements of the regime established by the original Superfund statute, the Comprehensive Environmental Response, Compensation, and Liability Act of 1980 (CERCLA). First, the courts apply joint-and-several liability when the harm is indivisible. Second, a right of contribution exists among defendants found jointly-and-severally liable. Third, in contribution actions, the relevant shares are determined by reference to comparative fault; for generators, this determination is typically made on the basis of the amount of waste generated. Fourth, following a settlement, the plaintiff's claim against the nonsettling defendants is reduced by the amount of the settlement (a *pro tanto setoff rule*). Fifth, a settling defendant is protected from any contribution actions. Sixth, a settling defendant can bring contribution actions against nonsettling defendants. Seventh, there is no detailed judicial supervision of the substantive adequacy of settlements. Eighth, in general, the claims involving the joint tort-feasors, including contribution claims, are sometimes litigated together in a single proceeding, although other times the contribution actions are severed and stayed (that is, tried after the plaintiff's case against the original defendants); we model the former approach.

This regime applies when the plaintiff is either EPA or a state, rather than a private party. The first and seventh elements have been fashioned by the courts as federal common law; the second through sixth are specified in SARA, and the eighth stems from the application of the Federal Rules of Civil Procedure. In contrast, when the plaintiff is a private party, the statute does not define the fourth through sixth elements, and some courts have fashioned federal common law rules that are different from the statutory rules that apply when EPA or a state is the plaintiff.

A regime of nonjoint liability can be defined, more straightforwardly, by reference to only two elements. In performing our comparison of the relative settlement-inducing effects, we shall first assume that, under nonjoint liability, the plaintiff's claim against each defendant is equal to that defendant's comparative fault. Thus, under nonjoint liability, EPA can recover the full cleanup costs only if it can locate all the defendants and if none of them are insolvent. Second, if the plaintiff settles with one defendant for more than that defendant's apportioned share of the liability, it can nonetheless recover the apportioned share of the other defendants. In this way, it is possible for the plaintiff to recover more than its damages. In contrast, under joint-and-several liability coupled with a pro tanto setoff rule, the plaintiff can never recover more than its damages when it settles with some defendants and litigates against others.

To perform the comparison between joint-and-several liability, on the one hand, and nonjoint liability, on the other, we specify a simple model, under which EPA has a claim of $100 against two defendants, Row and Column, each equally at fault. All the parties are risk neutral. We assume initially that the defendants are sufficiently solvent that they can satisfy the plaintiff's judgment. In a later section of this discussion, we consider the effects of limited solvency.

The probability that the plaintiff will prevail against each defendant is 50%. All the parties have accurate information about this value. We assume initially that the costs of litigation are zero and then examine how the results are affected by the presence of litigation costs.

With respect to the relationship between the plaintiff's *probabilities of success* in litigation against the two defendants, we consider two polar situations. In one, these probabilities are *independent*. Thus, the plaintiff's probability of success against one defendant is 50% regardless of whether the plaintiff has prevailed against, lost to, or settled with, the other defendant. For example, if the issue in a case is whether two separate defendants sent hazardous wastes to a site, EPA's probabilities of success will be independent. The fact that one defendant sent hazardous wastes makes it no more or less likely that the other did so as well.

In the other situation, the probabilities of success in litigation are *perfectly correlated*. Thus, if the plaintiff litigates against both defendants, it either prevails against both (with a probability of 50%) or loses against both (also with probability of 50%). For example, if the defendants argue that EPA's costs were inconsistent with the national contingency plan, the plaintiff's probabilities of success are perfectly correlated: if the defendants establish lack of consistency they will both prevail; otherwise, they will both lose.

The parties may either litigate or settle the claim. Settlement negotiations have the following structure. The plaintiff makes settlement offers to the two defendants. Row and Column decide simultaneously whether to accept these offers. We assume that costs of coordinating their actions are sufficiently high that they act noncooperatively. The plaintiff then litigates against the nonsettling defendants, if any. We adopt the convention that, if a party is indifferent between settlement and litigation, it settles.

The central conclusion of our analysis is that the comparison of the settlement-inducing properties of joint-and-several liability and nonjoint liability depends critically on the correlation of the plaintiff's probabilities of success. When these probabilities of success are independent, joint-and-several liability unambiguously discourages settlements, relative to nonjoint liability. When, in contrast, these probabilities are perfectly correlated, joint-and-several liability has a more complex effect: it encourages settlement when the litigation costs are low but may discourage settlements when these costs are high. A more detailed treatment of these issues can be found in Kornhauser and Revesz 1993, 1994a, 1994b.

We first examine the choice between settlement and litigation under nonjoint liability. We then perform the same analysis under joint-and-several liability for zero litigation costs and full solvency. Finally, we study for joint-and-several liability the effects of positive litigation costs and limited solvency.

Nonjoint Liability

The analysis of the choice between settlement and litigation under nonjoint liability is straightforward. The plaintiff's expected recovery from litigation is $50: it has a 50% probability of obtaining $50 from each defendant; each defendant's expected loss is therefore $25. Absent litigation costs, the plaintiff and the defendants are indifferent between litigation and settlement. For any level of litigation costs, settlement becomes preferable. For example, if each party's litigation costs were $5 (for the plaintiff, $5 against each of the defendants), the plaintiff's expected recovery from litigation would be only $20 from each defendant and each defendant's expected loss would be $30. The plaintiff and each defendant would prefer any settlement between $20 and $30 to litigation.

The result that under nonjoint liability the parties are indifferent between settlement and litigation in the absence of litigation costs and prefer to settle for any level of litigation costs does not change if the defendants have limited solvency. Say, for example, that Row's solvency is only $20. Then, in the absence of litigation costs, the plaintiff and Row are indifferent between litigation and a settlement for the plaintiff's expected recovery of $10 (a 50% probability of recovering Row's solvency of $20). For any level of litigation costs, the parties prefer to settle. Thus, while limited solvency affects the expected value of the plaintiff's claim as well as the amount at which the case would settle, it does not affect the choice between settlement and litigation.

Joint-and-Several Liability

As discussed above, the likelihood and nature of settlement under joint-and-several liability is determined by the correlation of the plaintiff's probabilities of success against the defendants. We consider the cases of independent and perfectly correlated probabilities.

Independent Probabilities. As a consequence of joint-and-several liability, the plaintiff recovers its full damages not only if it prevails against both defendants but also if it prevails against one and loses against the other. When the plaintiff's probabilities of success against the two defendants are independent, each of four different scenarios carries a probability of 25%: that the plaintiff prevails against both defendants, that the plaintiff prevails against Row and loses against Column, that the plaintiff prevails against Column and loses against Row, and that the plaintiff loses against both defendants. In the first three cases, carrying an aggregate probability of 75%, the plaintiff recovers its full damages of $100. Thus, its expected recovery from litigating with both defendants is $75.

In turn, each defendant's expected loss is $37.50. Row has a 50% probability of prevailing, and, therefore, of not having to pay anything. There is a probability of 25% that Row will lose and Column will win. In this case, Row has to pay the plaintiff's full damages of $100. Finally, there is a probability of 25% that Row and Column will both lose. Row then has to pay its share of $50, with Column paying the rest.

A risk-neutral plaintiff will not accept a settlement with both defendants that yields less than $75, but would find acceptable an aggregate settlement for $75 or more. What would happen if the plaintiff made settlement offers to the two defendants for $37.50 each, so that its aggregate recovery was equal to the expected recovery of litigating against both defendants? If one defendant, say Row, accepted the offer, would the other defendant accept it as well? Column would accept the settlement only if its expected loss from litigation is at least $37.50. Under the pro tanto setoff rule, Column's exposure in the event of litigation is reduced to $62.50: the plaintiff's damages of $100 minus Row's settlement of $37.50. But Column faces only a 50% probability of losing the litigation. Thus, in light of Row's settlement, its expected loss from litigation is only $31.25.

It therefore follows that if the plaintiff were to make offers of $37.50 to each defendant, at least one of them would reject the offer. The plaintiff's expected recovery would then be $68.75 (Row's settlement of $37.50 plus an expected recovery of $31.25 from litigating against Column). This amount is lower than the plaintiff's expected recovery from litigating against both defendants. Thus, the plaintiff would never make offers of $37.50 to each defendant. Similar logic establishes that no other pair of offers would give the plaintiff an expected recovery of at least $75 and yet be acceptable to the two defendants. Also, there is no scenario under which the plaintiff would receive an expected recovery of at least $75 by settling with one defendant and litigating against the other.

The preceding result would hold even if the two defendants were not equally at fault or if the plaintiff's probability of success were not 50%. It is still the case that the plaintiff would litigate rather than settle. This preference has two sources: the surplus that the plaintiff obtains from litigation as a result of joint-and-several liability when its probabilities of success against the defendants are independent, and the benefit that a nonsettling defendant receives from the setoff created by the plaintiff's settlement with the other defendant.

If the plaintiff were litigating against only one defendant rather than two, its expected recovery from litigation would be $50 rather than $75: it would have a 50% probability of recovering from that defendant its full damages of $100. Similarly, as we have indicated, if the plaintiff were litigating against two defendants under nonjoint liability, its expected recovery would also be $50: it has a 50% probability of recovering $50 from each of the defendants. Finally, if the plaintiff were litigating against two defendants under joint-and-several liability but its probabilities of success against the defendants were perfectly correlated, it would also have an expected recovery of only $50. The perfect correlation of the probabilities implies that the plaintiff either prevails against both defendants or loses against both; in the former case, which carries a probability of 50%, it recovers its full damages of $100, whereas in the latter case it gets nothing.

As a result of the surplus that the plaintiff obtains from litigating under joint-and-several liability when the probabilities of prevailing are independent, the plaintiff will not accept from one defendant a settlement that is too low even if it intends to litigate against the other. Say, for example, that the plaintiff accepted a settlement of $0 from Row and litigated against Column. Its expected recovery would then be only $50 (a 50% probability of recovering $100); the settlement with Row will have reduced its expected recovery by $25. If the plaintiff accepted a settlement of $10 from Row, its expected recovery from litigating with Column would be $45 (a 50% probability of recovering $90), for a total expected recovery of $55; the loss from the low settlement with Row is $20.

So as not to lose its surplus, the plaintiff would thus have to demand a sufficiently high settlement from Row. But a settlement that is sufficiently desirable for the plaintiff to accept confers a benefit upon Column. If, for example, the plaintiff were to settle with Row for $25, Column's expected loss from litigation would be $37.50—the same expected loss as if Row litigated. Any higher settlement with Row, reduces Column's expected loss. We have already shown that a settlement with Row for $37.50 reduces Column's expected loss from $37.50 to $31.25, giving it a benefit of $6.25. In order to recover $75, the plaintiff would have to obtain from Row a settlement of $50 (which would leave an expected recovery from Column of $25 and confer upon Column a benefit of $12.50). Row, however, would not agree to such a settlement because, given that Column litigates, it is better off litigating as well and facing an expected loss of only $37.50.

We have thus illustrated why the plaintiff cannot capture the full benefit of Row's settlement if its probabilities of success are independent. Part of this settlement confers an external benefit upon Column. It is this externality that stands in the way of settlement. Indeed, the only way that the plaintiff can obtain the full benefit of a defendant's payment is by litigating, because if it settles, part of the benefit accrues to the other defendant, reducing the plaintiff's expected recovery from litigation.

Perfectly Correlated Probabilities. The problem changes considerably when the plaintiff's probabilities of success against both defendants are perfectly correlated. If the plaintiff litigates against both defendants, it either prevails against both (with a probability of 50%) or loses against both (also with a probability of 50%). Its expected recovery from litigation is $50 rather than $75; each defendant's expected loss is then $25.

In the case of perfectly correlated probabilities, the plaintiff will settle with both defendants. It is easy to see that the plaintiff will settle with at least one of the defendants. Say that the plaintiff settles with Row for $10: it faces a 50% probability of recovering $90 from Column, and its total expected recovery is $55—$5 higher than its recovery from litigating against both defendants. The effect of this settlement is to give the plaintiff $10 with certainty, but reduce its expected recovery from litigation by $5. As a result, settlement with one defendant and litigation against the other is always more attractive to the plaintiff than litigation against both defendants.

It is also easy to show that, for the example that we are analyzing, the plaintiff in fact settles with both defendants, for $25 and $37.50, respectively. Given that Row settles for $25, Column's expected loss through litigation is $37.50 (a 50% probability of paying the plaintiff's damages of $100 minus Row's settlement of $25), and would therefore accept a settlement for that amount. Moreover, given that Column settles for $37.50, Row's expected loss through litigation is $31.25 (a 50% probability of paying the plaintiff's damages of $100 minus Column's settlement of $37.50), and therefore Row would prefer to settle for $25. The same argument establishes that the plaintiff would be no better off settling with one defendant and litigating against the other.

We show elsewhere that, for perfectly correlated probabilities, the plaintiff settles with both defendants if their shares of the liability are sufficiently similar and settles with one defendant—the one with the larger share of the liability—and litigates against the other if the defendant's shares of the liability are sufficiently different (Kornhauser and Revesz 1993).

The Effects of Litigation Costs

So far, for analytical clarity, we have dealt with the case in which litigation costs are zero. Litigation costs always have the effect of making settlement relatively more attractive because the plaintiff and the defendants can save these outlays and divide them among each other in some fashion if they choose not to litigate. The question is whether the presence of litigation costs affects the relative settlement-inducing properties of joint-and-several liability and nonjoint liability.

When litigation costs are sufficiently low, the conclusions derived in the prior section (on the choice between settlement and litigation in joint-and-several liability situations) remain unchanged. When the plaintiff's probabilities of success are independent, the savings that can be realized by eliminating litigation costs will be insufficient to eliminate the effects of the externality that stands in the way of settlements. When the plaintiff's probabilities of success are perfectly correlated, joint-and-several liability will retain its settlement-inducing advantage for litigation costs below a given threshold.

When litigation costs are sufficiently high, however, the analysis is different. One feature of the pro tanto setoff, as we have already explained, is that when the plaintiff settles with one defendant and litigates against the other, it cannot recover more than its full damages. Under nonjoint liability, in contrast, the plaintiff's recovery is not constrained in this manner. When the plaintiff makes a take-it-or-leave-it offer and litigation costs are sufficiently high, a defendant under nonjoint liability will be willing to pay more than its apportioned share of the liability in order to avoid litigation. For this level of litigation costs, the plaintiff obtains larger settlements under nonjoint liability than under joint-and-several liability.

At the same time, however, the plaintiff's recovery from litigating against both defendants is higher under joint-and-several liability than under nonjoint liability when the plaintiff's probabilities of success are independent and is equal under both rules when the probabilities are perfectly correlated. If a defendant were able to make take-it-or-leave-it offers, it would offer the plaintiff the smallest amount that would make the plaintiff indifferent between settling and litigating. The amount thus offered is higher under joint-and-several liability than under nonjoint liability when the plaintiff's probabilities of success are independent and is equal under both rules when the probabilities are perfectly correlated.

In the real world, none of the parties can make take-it-or-leave-it offers; instead, the parties must engage in bargaining. One can think of the case in which the plaintiff makes a take-it-or-leave-it offer as defining the upper bound of the settlement range: the plaintiff captures the full surplus of settlement. Similarly, one can think of the case in which one of the defendants makes a take-it-or-leave-it offer as defining the lower bound of the settlement range: here, the defendant captures the full surplus of settlement. For sufficiently high litigation costs, the settlement range is greater under nonjoint liability. It is true that in this situation, risk-neutral parties with accurate information about the plaintiff's probabilities of success will settle under both joint-and-several liability and nonjoint liability.

In contrast, sufficiently optimistic parties will litigate under both rules. The higher settlement range for nonjoint liability implies that for an intermediate range of optimism, there will be settlements under nonjoint liability but not under joint-and-several liability.

In summary, joint-and-several liability deters settlements when the plaintiff's probabilities of success are independent not only when litigation costs are low but also when they are high. In contrast, when the plaintiff's probabilities of success are perfectly correlated, joint-and-several liability promotes settlements when litigation costs are low, but deters settlements when they are high. These results of this part of our analysis are summarized in Table 3.

Table 3. Effects of Joint-and-Several Liability on Settlements under Different Levels of Litigation Costs (High Solvencies) Relative to Nonjoint Liability

	Low litigation costs	High litigation costs
Independent probabilities	Discourages settlement	Discourages settlement
Perfectly correlated probabilities	Encourages settlement	Discourages settlement

The Effects of Limited Solvency

As we indicated previously, the limited solvency of the defendants does not affect the choice between settlement and litigation under nonjoint liability. The situation is different under joint-and-several liability. We consider first how limited solvency would affect the choice between settlement and litigation if the plaintiff's probabilities of success are independent (we assume for this discussion that litigation costs are zero). If one of the defendants, say Row, has limited solvency, the plaintiff nonetheless litigates against both defendants if this solvency is above a threshold. For example, if Row's solvency is $80 and the plaintiff litigates against both defendants, its expected recovery is $37.50 from Column but only $32.50 from Row (with a probability of 25%, the plaintiff prevails against both defendants and recovers $50 from Row, and, also with a probability of 25%, the plaintiff prevails only against Row and recovers Row's solvency of $80 rather than its full damages of $100). In contrast, if the plaintiff settled with Column for $37.50, Row's expected loss from litigation, and consequently the maximum settlement that it would offer, would be only $31.25 (a 50% probability of paying the plaintiff's damages of $100 minus Column's settlement of $37.50).

When Row's solvency is sufficiently low, however, the plaintiff settles with both defendants. Consider the case in which Row's solvency is $40. If the plaintiff litigates against both defendants, its expected recovery is $60 (with a probability of 25%, it prevails only against Column and recovers $100; with a probability of 25%, it prevails against both and recovers $40 from Row and $60 from Column; and with a probability of 25%, it prevails only against Row and recovers $40). In turn, Row's expected loss is $20 and Column's expected loss is $40.

If the plaintiff offered Row a settlement of $20, its expected recovery from Column is $40 (a 50% probability of recovering its damages of $100 minus Row's settlement of $20), and Column would be willing to settle for this amount. In turn, if the plaintiff offered Column a settlement of $40, its expected recovery from Row is $20 (a 50% probability of recovering its solvency of $40), and Row would be willing to settle for this amount. Thus, as in the case of nonjoint liability, when the solvency of one of the defendants is sufficiently low and litigation costs are zero, the parties are indifferent between settling and litigating.

In summary, the result that joint-and-several liability discourages settlements when the plaintiff's probabilities of success are independent holds over a range of solvencies. A similar analysis (see Kornhauser and Revesz 1994b) establishes that, when the plaintiff's probabilities of success are perfectly correlated, joint-and-several liability promotes settlements over a range of solvencies. For solvencies below a given threshold, however, joint-and-several liability has the same settling-inducing properties as nonjoint liability. The relevant results of this part of our analysis are summarized in Table 4.

Implications of the Analysis

The preceding analysis illustrates the difficulties of making categorical statements about the settlement-inducing properties of joint-and-several liability. Even under the simple formulation of our model, whether joint-and-several liability promotes or deters settlements depends crucially upon three factors: the degree of correlation of the plaintiff's probabilities of success, the level of litigation costs, and the solvencies of the defendants.

Of course, an extensive empirical study might reveal that, for the Superfund scheme as a whole, joint-and-several liability is having one effect rather than the other. Relevant questions would include the extent to which the plaintiff's probabilities of success are correlated, the relative levels of litigation costs, and the solvencies of PRPs. But for now, given that such an inquiry has not taken place, one ought to be agnostic about the settlement-inducing properties of joint-and-several liability.

One might complain that we have dealt only with a highly stylized model that makes quite restrictive assumptions. Elsewhere (Kornhauser and Revesz 1993), we have shown that our central results persist even when many of the assumptions are relaxed. We are quite confident that there is no realistic scenario under which joint-and-several liability would either always promote or always deter settlements.

Table 4. Effects of Joint-and-Several Liability on Settlements under Different Levels of Solvency (Low Litigation Costs) Relative to Nonjoint Liability

	High solvency	Low solvency
Independent probabilities	Discourages settlement	Neutral effect
Perfectly correlated probabilities	Encourages settlement	Neutral effect

FAIRNESS IN ALLOCATING LIABILITY

While opponents of joint-and-several liability in the Superfund context have complained about the unfairness of joint-and-several liability, they have not attempted to define with care the yardstick against which they measure fairness (see National Commission on Superfund 1994, 16–20). We believe that the most plausible fairness objective is that each generator should pay for its share of the harm. (The allocation between generators and site owners raises more difficult issues that we do not address here.)

This fairness objective is compromised not only when a generator pays more than its share of the harm, but also when the converse is true. If a generator pays less than its share, either another generator must cover the shortfall, or EPA is not going to recover the full cost of the cleanup. The unfunded portion would then be financed through the Hazardous Substance Superfund (the CERCLA-established trust fund)—three separate taxes levied on chemicals, petroleum products, and general corporate profits. An assessment of fairness then depends on the fairness of the tax measures.

Four Fairness Issues

The comparison of the relative fairness of joint-and-several liability and nonjoint liability raises four principal issues. Three fairness issues arise when the defendants are fully solvent: the size of the plaintiff's expected recovery when it litigates against the defendants; the division of the plaintiff's recovery among litigating defendants; and the effects of settlements. A fourth issue arises when the defendants have limited solvency: the division of the burden of insolvency between the plaintiff and the solvent defendant.

A question relevant to all four issues is whether one should assess fairness ex ante (in terms of the parties' expected payments) or ex post (in terms of the actual payments in particular cases). We largely confine our remarks here to ex ante assessments.

Size of the Plaintiff's Recovery. First, as indicated in the previous discussion on settlement-inducing properties, except when the plaintiff's probabilities of success against the defendants are perfectly correlated, joint-and-several liability leads to a higher expected recovery than nonjoint liability. Recall the example in which the plaintiff's damages are $100 and its probabilities of success against each of the defendants are 50%, and the defendants are equally at fault and fully solvent. The plaintiff's expected recovery is $50 under nonjoint liability, $50 under joint-and-several liability when the plaintiff's probabilities of success are perfectly correlated, and $75 under joint-and-several liability when the plaintiff's probabilities of success are independent. (In the range between independence and perfect correlation, the plaintiff's recovery is between $50 and $75.)

Thus, except when the plaintiff's probabilities of success are perfectly correlated, an effect of joint-and-several liability is to transfer resources from the defendants to the plaintiff. The fairness consequence of this transfer depends upon why the plaintiff's probability of success against each of the defendants is only 50%. It could be that the defendants are in fact liable but that the plaintiff has difficulty in proving their liability. In this case, joint-and-several liability is attractive on fairness grounds because it brings a defendant's expected liability closer into line with the harm that it caused.

Alternatively, it could be that there is true uncertainty about whether the defendants are liable, and that this uncertainty is captured by the 50% probability. Then, joint-and-several liability is undesirable because it increases a defendant's expected liability beyond the level of the harm the defendant caused.

Division of the Plaintiff's Recovery. The second issue concerns the allocation of expected liability among litigating defendants. From this perspective, joint-and-several liability performs badly: it places a disproportionate burden on the defendant with the smaller share of the liability, except when the plaintiff's probabilities of success are perfectly correlated.

Consider an example in which, instead of being equally at fault, Row and Column are 25% and 75% at fault, respectively; the plaintiff's probabilities of prevailing against each of the defendants remains at 50%, and these probabilities are independent. There are then four possible scenarios, each carrying a probability of 25%:
- The plaintiff prevails against both defendants and collects $25 from Row and $75 from Column.
- The plaintiff prevails against Row and loses to Column, and collects $100 from Row.
- The plaintiff loses to Row and prevails against Column, and collects $100 from Column.
- The plaintiff loses to both defendants and does not recover anything.

Thus, Row pays $25 with probability 25% and $100 with probability 25%; its expected liability is then $31.25. In turn, Column pays $75 with probability 25% and $100 with probability 25%, and its expected liability is $42.75. Thus, while Row's contribution to the harm is only one-third that of Column's, its expected liability is about three-quarters that of Column's.

The preceding example shows that this disproportionate effect stems exclusively from the fact that under joint-and-several liability the plaintiff might prevail against the defendant with the lower responsibility for the harm but lose against the other defendant, and that the defendant with the lower responsibility is then required to pay the plaintiff's full damages. In contrast, under nonjoint liability (and under joint-and-several liability when the plaintiff's probabilities of success are perfectly correlated), each defendant's expected liability is proportional to its responsibility for the harm.

The Effects of Settlements. The possibility of settlements introduces the third fairness issue, that of the effects of settlements. This possibility places (as did the issue of the division of a plaintiff's recovery) a disproportionate burden on the defendant with the smaller share of the liability. Indeed, for the legal regime that is analyzed in this chapter's section on settlement-inducing properties, and which employs a pro tanto setoff rule, each defendant settles for the same amount, even when their shares of the harm are different. Consider the example in which the litigation costs are sufficiently high that they induce the parties to settle, and in which the plaintiff makes take-it-or-leave-it offers to the defendants.

The largest settlement that Row will accept, S_R, conditional on Column settling for S_C (which is less than the plaintiff's damages D), is given by

$$S_R = p(D - S_C) + t$$

where p is the plaintiff's probability of success against each defendant, and t is each defendant's litigation costs. Similarly, the largest settlement that Row will accept, S_C, conditional on Column's settling for S_R (which is less than the plaintiff's damages D) is given by

$$S_C = p(D - S_C) + t$$

Thus,

$$S_R = S_C = (Dp + t)/(1 + p)$$

Thus, when litigation costs are sufficiently high that the parties settle despite the independence of the plaintiff's probabilities of success, the plaintiff extracts from each defendant an equal settlement, regardless of the differences in the defendants' shares of the harm.[3]

In contrast, recall that under nonjoint liability, each defendant's expected liability is proportional to its responsibility for the harm. The plaintiff, if it made take-it-or-leave-it offers, could extract from each defendant in settlement this amount plus the defendant's litigation costs. If each defendant's litigation costs are independent of their share of the liability, the defendant with the smaller share will pay a disproportionate amount, but it will be less disproportionate than what it would have paid under joint-and-several liability.

Division of the Burden of Insolvency. The fourth fairness issue arises if one of the defendants has limited solvency. Our assessment of fairness here is neither fully ex ante nor fully ex post. A fully ex ante perspective would consider the likelihood that each defendant would become insolvent; instead our discussion assumes that one defendant is already insolvent. On the other hand, our discussion is not fully ex post because we assess fairness in terms of expected litigation (and settlement) outcomes. We hope to consider fairness issues more fully in a subsequent article.

We have studied elsewhere how the shortfall caused by the limited solvency of one defendant is allocated between the plaintiff and the remaining solvent defendant under joint-and-several liability (Kornhauser and Revesz 1994b). That study revealed that, over a broad range of solvencies, and for less-than-perfectly correlated probabilities, the plaintiff bears the full shortfall, and it is never the case that the full shortfall is borne by the solvent defendant. This conclusion challenges the accepted wisdom that, under joint-and-several liability, the burden of one defendant's insolvency falls exclusively on its codefendants (Sugarman 1992, 1188).

The reason for the entrenchment of this erroneous view may be that judges and commentators implicitly consider only the situation in which the plaintiff's probabilities of success are perfectly correlated and the plaintiff litigates against both defendants. Then, any shortfall caused by one defendant's limited solvency is borne by the other defendant. If, however, the correlation of the probabilities is less than perfect, the plaintiff's expected recovery is reduced because it might prevail only against the defendant with limited solvency. Moreover, the focus on litigation overlooks two important facts: the plaintiff often is better off settling, and the amount that it can recover in settlement is a function of the solvency of the defendants.

Nonetheless, under joint-and-several liability, the shortfall caused by one defendant's limited solvency is generally shared between the solvent defendant and the plaintiff. In contrast, under nonjoint liability, the full shortfall is borne by the plaintiff. Thus, nonjoint liability puts additional pressure on the Hazardous Substance Superfund. Which is a fairer way of meeting the shortfall? The full answer to this question would have to include an analysis of the distribution of orphan shares at Superfund sites as well as of the incidence of the taxes needed to fund the Superfund (see Chapter 6 of this book and Probst and Portney 1992, 16)—an inquiry that we cannot undertake here.

In summary, joint-and-several liability performs worse in terms of fairly allocating liability among defendants but does not necessarily perform worse in terms of fairly allocating liability between the plaintiff, on the one hand, and the defendants, on the other.

CONCLUSION

Our analysis reveals that along three important dimensions—incentives for waste reduction and care, incentives for settlement, and fairness—one cannot make categorical claims about the relative desirability of joint-and-several liability and nonjoint liability. Neither rule performs consistently better than the other.

We do, however, identify the factors that determine which of the rules is more desirable in particular situations. These factors include the solvency of the defendants, the correlation of the plaintiff's probabilities of success, the relative level of litigation costs, and the incidence of the taxes used to cover unfunded liabilities. Extensive empirical work concerning these factors (for example, attempting to estimate the correlation of the plaintiff's probabilities of success at a typical Superfund site) might suggest that, on the whole, one of the liability rules is better than the other. Such work, however, has not been done and will not be done in time to inform the reauthorization debate; moreover, our analysis provides a framework for the performance of such empirical work. If the burden is to be placed on those urging departure from the status quo, as it probably should be, the proponents of abandoning joint-and-several liability have fallen short.

We end by underscoring that our study has compared joint-and-several liability with nonjoint liability on the basis of only three criteria. There are, of course other relevant criteria, such as the effects of these alternative rules on the availability of insurance or on the level of litigation costs.

With respect to insurance availability, joint-and-several liability is problematic because the probability of liability is determined by the actions of nonpolicyholders, whom the insurer cannot identify in advance; thus, it is thought to significantly decrease the availability of insurance (Abraham 1988, 959–60).

The comparison on the basis of litigation costs is less straightforward. On the one hand, joint-and-several liability raises the stakes whenever orphan shares must be allocated to solvent parties: if litigation costs rise with the stakes of the litigation, joint-and-several liability will increase the level of these costs. On the other hand, as we have shown above, under some circumstances joint-and-several liability promotes settlements, thereby reducing litigation costs.

ACKNOWLEDGMENTS

This work was funded by a grant from EPA's Office of Exploratory Research (contract #818460-01-1). We also acknowledge the generous financial support of the Filomen D'Agostino Greenberg and Max E. Greenberg Research Fund at the New York University School of Law. Vicki Been, Colleen Shannon, and Richard Stewart gave us valuable comments. The views presented in this chapter are those of the authors and not those of EPA.

ENDNOTES

[1] Traditionally, the academic literature distinguishes between *activity levels* and *levels of care*. The amount of wastes generated would typically be thought of as an activity level. The two concepts, though, are analytically analogous. It may be, however, that the legal system does not have much experience administering a negligence rule triggered by activity levels. This distinction, however, in no way affects the illustration of the comparative effects of joint-and-several liability and nonjoint liability presented in the text.

[2] This discussion assumes that the liability of the defendants is proportional to the amount of wastes generated. If, instead, the defendants could be charged for the marginal cost of their actions, this problem would not arise. In this example, generating 6 units rather than 5 would cost Column an additional $21 under the marginal cost approach—the full increase in total damages. Such a rule, however, is difficult to administer, because it requires courts to make determinations about marginal costs, and it is not used by the courts.

[3] Of course, this feature makes settlement relatively less attractive to the defendant with the smaller share of the liability. In the Superfund context, EPA generally makes settlement offers that are proportional to a PRP's volumetric contribution, even though EPA's recovery would be greater if it made disproportionate offers.

REFERENCES

Abraham, Kenneth S. 1988. Environmental Liability and the Limits of Insurance. *Columbia Law Review* 88: 942–88.

Kornhauser, Lewis A. and Richard L. Revesz. 1989. Sharing Damages among Multiple Tortfeasors. *Yale Law Journal* 98: 831–84.

———. 1990. Apportioning Damages among Potentially Insolvent Actors. *Journal of Legal Studies* 19: 617–51.

———. 1993. Settlements under Joint and Several Liability. *New York University Law Review* 68: 427–93.

———. 1994a. Multidefendant Settlements: The Impact of Joint and Several Liability. *Journal of Legal Studies* 23: 41–76.

———. 1994b. Multidefendant Settlements under Joint and Several Liability: The Problem of Insolvency. *Journal of Legal Studies* 23: 517–42.

National Commission on Superfund. 1994. *Final Consensus Report of the National Commission on Superfund.*

Probst, Katherine N. and Paul R. Portney. 1992. *Assigning Liability for Superfund Cleanups.* Washington, D.C.: Resources for the Future.

Sugarman, Stephen D. 1992. A Restatement of Torts. *Stanford Law Review* 44: 1163–1208.

6

Evaluating the Impact of Alternative Superfund Financing Schemes

Katherine N. Probst

The Superfund debate thus far has focused on the strengths and weaknesses of the current liability standards and, to a much lesser extent, of alternative approaches (Probst and Portney 1991). A subject that gets much less attention is the question of who will *pay* for the increased trust fund that would be necessitated by any revision to the current liability scheme and who will *benefit* by changes to the liability scheme. Any *decrease* in liability requires a corollary *increase* in one or more of the current Superfund taxes, or the levy of a new tax.

Any evaluation of "who pays" for Superfund cleanups requires an assessment of who pays under the liability scheme and who is paying the Superfund taxes. Needless to say, the first task is far more difficult than the second. This is due in large part to the fact that the law does not require responsible parties (RPs) to report to the U.S. Environmental Protection Agency (EPA) how much they are actually spending on site studies and cleanup. Without information on the expenditures of RPs at each site and on how much the individual RPs are paying for cleanups, it is difficult to know how much is actually being spent, much less which companies and sectors of the economy are bearing the burden.

Very little is known about who bears the brunt of cleanup costs under the current liability scheme, let alone how the distribution of these costs would change under alternative liability approaches. Yet, this information is of critical importance to any evaluation of the economic impact of Superfund liability, as well as to discussions of what kind of tax mechanism should be used to raise additional trust fund revenues should the liability scheme be revised. (The trust fund, formally known as the Hazardous Substance Response Fund, may be used by EPA to finance cleanups.) Each liability alternative releases a different set of RPs from liability. Thus, some industries may well benefit more from one alternative than another.

Ascertaining who pays under the current Superfund program is not an easy task. Little information is publicly available on how much is being spent to clean up National Priorities List (NPL) sites, much less who is footing the bill. This lack of information is complicated by the confidentiality of the terms of most settlement agreements, making it impossible to estimate the distribution of costs among multiple parties at an individual site. In addition, those who bear the initial costs of cleanup may not be those who ultimately absorb the financial liability. In many cases, costs are reallocated after site studies and cleanups are completed. This reallocation can happen in two ways: the government may pay for site studies or cleanup and then seek to recover its costs from RPs, or one or a few RPs may pay for cleanup and seek reimbursement from nonsettling RPs and/or from their insurers. In addition, to the extent that insurers are found to be liable under their general liability policies, they will most likely obtain reimbursement from reinsurers. For all these reasons, then, it is not possible to ascertain how cleanup costs will ultimately be distributed. However, it *is* possible to develop ballpark estimates of the likely initial distribution of cleanup costs among different sectors of the economy, based on the type of site and the cause of contamination, the subject of this chapter.

To examine the pros and cons of the current financing scheme and the alternatives being proposed, it is important to look at the combination of the liability and tax mechanisms, as well as at the broader effects on the national economy. In conducting research to address this gap, we hope to shed light on several different questions relating to the economic impact of Superfund financing approaches.

- Which sectors of the economy are likely to pay the largest burden of cleanup costs under the current liability system?
- Which sectors are paying the largest percentage of trust fund taxes?
- Which sectors benefit, and which lose, under different alternative liability schemes?
- Is the effect of the current trust fund taxes noticeable, in terms of the economy as a whole?
- What is the likely impact of Superfund liability in the insurance sector? What kinds of costs (cleanup or transaction costs) are the largest burden on this industry?

METHODOLOGY: ESTIMATING WHO PAYS

To find answers to the above questions, I conducted research with colleagues from Resources for the Future (RFF), The Brookings Institution, and the University of Texas–Austin. Our research has three main components: the development of a database to allow us to assign liability to different industry sectors under both the current liability scheme as well as under alternative approaches; the use of an input-output model to assess the incidence of Superfund taxes; and, finally, a look at the short-term economic implications of both the liability and tax impacts on some of the key industries affected by Superfund. This chapter focuses on the impacts of alternative liability schemes on the size of the trust fund and on key industry sectors. Our findings regarding our evaluation of alternative taxes and the economic implications of changes in liability on key industry sectors are described only briefly.

Our research requires developing information on who bears the initial costs under the present financing scheme in Superfund; that is, we estimate who is likely to pay for site cleanup, on a site-by-site basis, under the current liability approach. Ideally, site-specific information would be available on the costs of cleanup at each site, the percentage of these costs borne by the trust fund and by specific industry sectors, and actual transaction costs. This information does not exist, in large part because cleanup activities at most sites have not yet been completed and because information on actual cleanup costs at sites where RPs take the lead is not required to be reported to EPA. In order to make educated cost estimates in our analysis of the liability alternatives, we created the RFF NPL Database.

Our NPL database includes information on 1,134 nonfederal facility NPL sites. Much of the data is from an August 1993 survey (the "RPM Survey") conducted by EPA to collect information on all NPL sites from staff in EPA's ten regional offices (U.S. EPA 1993). Information from the RPM Survey is augmented by an earlier database we developed that integrates information from a number of other EPA databases. Each site in the database is assigned a cleanup cost, based on the type of facility on site (that is, chemical manufacturing facility, landfill, wood-preserving site, and so forth), and a likely share of total site costs devoted by RPs to transaction costs (based on the number of RPs at the site). In addition, we assign financial responsibility for each site to a specific industry, using Standard Industrial Classification codes where possible.

Cleanup costs are assigned to individual industries for each NPL site, based on our determination of which industries are most likely to be held financially liable. For example, we assume that the mining industry will bear the entire cost of cleaning up all mining sites on the NPL and that the chemical industry will bear the cost of cleaning up contamination at sites that are located at chemical manufacturing facilities. In this particular analysis, the impact of insurance contributions that would defray RP costs is not estimated.

For sites at which many firms and/or government agencies deposited wastes, or at which no major RP can be identified, assigning site costs to a specific industry is much more problematic. Over one-third of the NPL is comprised of sites where the wastes were deposited by many RPs, such as landfills, recycling facilities, and commercial waste handling and disposal sites (see Figure 1). We assign cleanup costs to a cross-section of industries for 230 of these sites, where some information is available on the industries that contributed waste to each site. For the remaining 147 sites, no information is available on the waste contributors. As a result, site costs are not assigned to any industry sector for this latter group of sites; in our estimates we simply label these costs "not attributed."[1]

Once we identify which industry is likely to bear the initial costs for cleanup at each NPL site, our research next turns to estimating the total cleanup costs to be borne by each industry sector. To develop this estimate, average site cleanup cost estimates are assigned for different categories of sites. These estimates are based on cost analyses by EPA, the University of Tennessee, and other sources (Russell, Colglazier, and English 1991). As shown in Table 1, there is a tremendous range in the

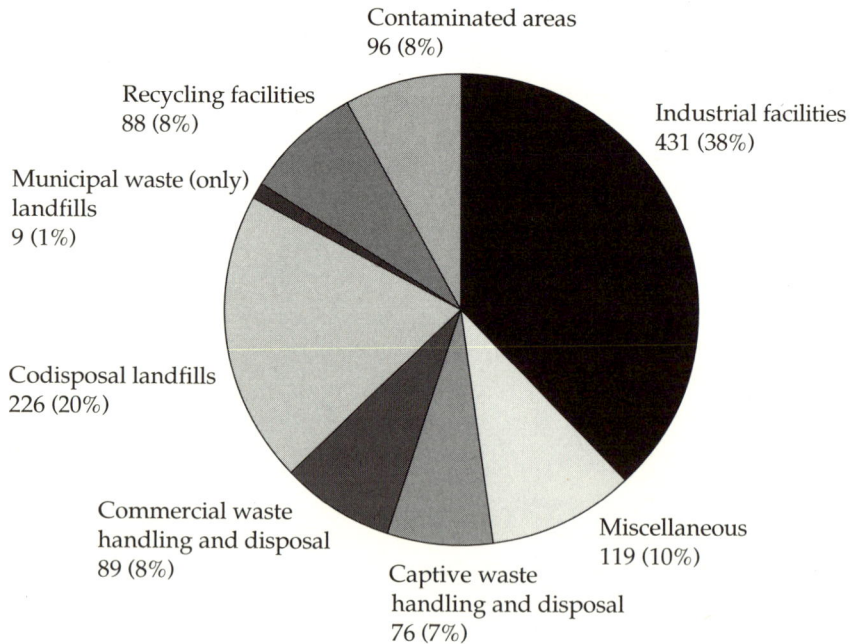

Figure 1. NPL Sites by Site Type (1,134 sites)
Source: RFF NPL Database 1994.

Table 1. Estimated Average Cleanup Costs by Type of Site

Site type	Estimated average cleanup cost ($ millions)
1. Asbestos	12.7
2. Chemical manufacturing	41.1
3. Drum recycling	18.9
4. Electrical	26.4
5. Landfill	29.0
6. Leaking container	34.4
7. Manufacturing	13.5
8. Metal working	13.0
9. Mining	170.4
10. Plating	14.0
11. Radiological tailings	75.4
12. Surface impoundment	24.9
13. Waste oil	32.3
14. Wellfield	14.9
15. Wood-preserving	40.6

Source: Author's calculations based on Colglazier, Cox, and Davis 1991, 45.

estimated average cleanup costs for different types of sites, from a high of $170.4 million for mining sites to a low of $12.7 million for asbestos sites.

A recent analysis of cleanups of Department of Energy (DOE) sites found dramatic cost-savings when the private sector conducts cleanups over what cleanups conducted by the federal government would cost.[2] To take this private-sector efficiency into account, we assume that the private sector is able to achieve a 20% cost-savings over the government for NPL site cleanups.[3] If all cleanups were implemented by RPs, the average cost of an NPL site cleanup would be $23.3 million, approximately $6 million (20%) less than the government average of $29.1 million.[4]

For each site where RPs pay for some or all of site cleanup costs, we estimate RP transaction costs. Lacking site-specific data on transaction costs, we assume that transaction costs are related to the number of RPs at the site and that the percentage of total site costs attributable to transaction costs increases with the number of RPs.[5] Our estimates of transaction costs range from 5% of total RP costs (that is, the sum of cleanup and transaction costs) for sites with one RP, to 30% of total costs for sites with more than 50 RPs (see Table 2). In addition to determining how each alternative affects *total* RP transaction costs, we also examine the distribution of transaction costs among industries.

Our data, shown in Figure 2, indicate that 59% of nonfederal facility NPL sites have fewer than eleven RPs, while 13% of sites have more than one hundred RPs.

Because we estimate transaction costs as a percentage of the total costs incurred by RPs, the transaction-cost share under each liability

Table 2. RP Transaction Costs Assumed

Number of RPs	Transaction costs (as percent of total costs)
Orphan	0
1 RP	5
2–10 RPs	20
11–50 RPs	25
More than 50 RPs	30
Unknown	15

Source: RFF NPL Database 1994.

option decreases in direct relationship to the percentage of costs paid for by RPs. That is, the more cleanups paid for by the trust fund, the lower the total RP transaction costs.

It is important to note that we quantify only the likely *reductions* in private-sector transaction costs to RPs for each liability alternative. However, under the two alternatives examined, some new transaction costs will likely be incurred. For example, if RPs are released from

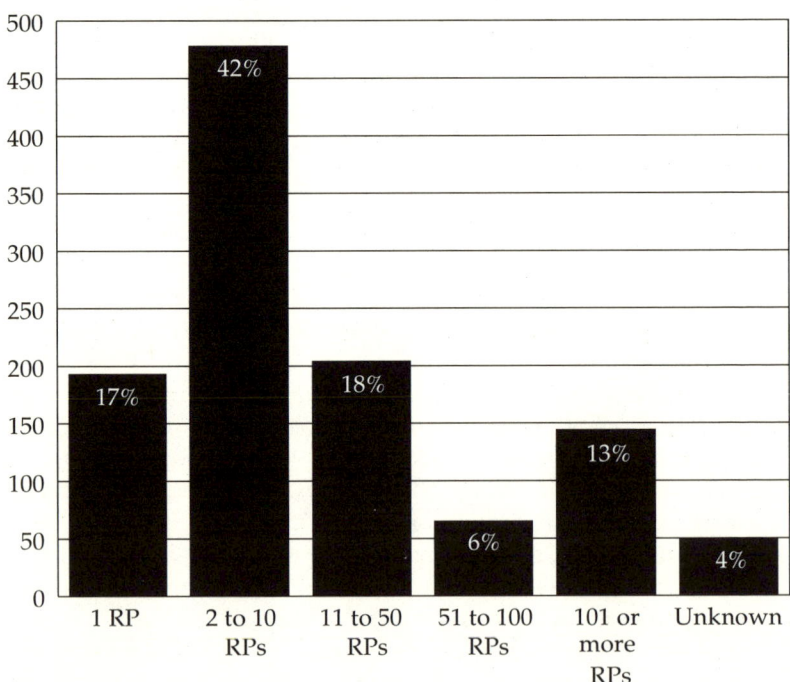

Figure 2. Estimated Distribution of NPL Sites by Number of RPs (1,134 sites)

Source: RFF NPL Database 1994.

Superfund liability for all hazardous substances disposed at multiparty sites before a specified date, at least some increased litigation is likely because some RPs argue that they really did dispose of "their" hazardous substances before the liability cutoff date. Similarly, some proposals retain liability for waste disposal that was in violation of the rules and regulations at the time. This is a much higher legal standard than is required under the current Superfund law and will likely lead to much higher transaction costs at sites.

It is important to note that the cost estimates presented are for *future* cleanup and transaction costs for cleanup of 1,134 nonfederal NPL sites. We confine cost estimates to the current NPL because our estimates of the cost and financial impact on key industries are based on site-specific data described earlier. With no way to collect similar data on future sites, we cannot predict the characteristics of sites to be added to the NPL in the future.

Because we are looking at sites currently on the NPL, we also take into account money spent to date on site studies and cleanup, both by the government (that is, trust fund dollars) and by RPs. Thus, the estimates presented in this chapter are for the *remaining* costs of cleaning up the 1,134 NPL sites in our database. According to EPA, $4.0 billion in trust fund moneys have been spent on site studies and cleanups at NPL sites as of the end of fiscal year 1993.[6] As noted earlier, data on what RPs have actually spent is not readily available. As a result, we make assumptions about the likely pace of RP expenditures to arrive at the estimate that RPs have spent $2.9 billion as of the end of fiscal year 1993 on site studies and cleanup at NPL sites.

ALTERNATIVE LIABILITY SCHEMES

Two major proposals dominated the 1994 liability debate: the Superfund reform bill proposed by the Clinton administration in February 1994 (H.R. 3800) and an eight-point plan, championed by the National Association for the Advancement of Colored People (NAACP) and supported by an unusual coalition, the Alliance for a Superfund Action Partnership (ASAP), which included major insurance companies, industrial corporations, small businesses, and local governments.[7]

The Clinton administration's H.R. 3800 proposal called for relatively modest changes in the current liability scheme and the creation of a new "environmental insurance resolution fund" (EIRF), to be financed by fees levied on insurance companies. The bill included provisions to exempt truly small contributors to a site (de micromis parties) from liability and expedited settlements with small contributors of hazardous substances at sites (de minimis parties). In addition, H.R. 3800 would have limited the liability of RPs that generate or transport municipal solid waste.[8] Perhaps the largest change would have been a softening of joint-and-several liability, as the administration's bill would have resulted in a move to a de facto proportional liability scheme in many cases. To accomplish this end, a neutral third party would have been responsible for allocating liability among RPs according to a set of specific factors. The trust fund would have covered the cleanup cost that otherwise would have been attributed to those RPs who were insolvent or could not be found, called the "orphan shares."

The EIRF, a product of negotiations among a small group of major industrial and insurance companies, was intended to "solve" the insurance problem. There is ample evidence that a large percentage of the transaction costs generated as a result of Superfund relate to disputes between insurers and those they insure over who should pay for cleanups (Acton and Dixon 1992). The EIRF would have provided a mechanism for resolving these disputes without litigation. Under H.R. 3800, the EIRF would have offered RPs who hold insurance policies financial settlements in connection with contamination caused by disposal of hazardous substances that took place on or before December 31, 1985. The EIRF would have reimbursed eligible RPs for 20%, 40%, or 60% of their costs, depending on the state in which the claim is brought.[9]

The eight-point plan called for by ASAP would have eliminated the application of retroactive liability for wastes disposed at multiparty sites before a certain date (which is not specified but would be some time between the enactment of CERCLA on December 11, 1980 and the present), except where that disposal violated laws in effect at the time. The proposal called for new taxes to create an increased trust fund of $3.0 to $4.6 billion annually. These taxes would come from a doubling of the corporate environmental tax and new taxes on insurers and small businesses.

This chapter examines three liability alternatives: the current Superfund program (referred to as Status Quo); the ASAP proposal using a 1980 cutoff date for eliminating retroactive liability at multiparty sites (referred to as ASAP Pre-1980); and the Clinton administration proposal (referred to as H.R. 3800).[10] The purpose here is to suggest the broad outlines of three liability options and to trace out the likely financial effects of these changes. A brief description of each liability option follows.

Option 1. The Current Superfund Program (Status Quo)

Under the current Superfund law, a wide variety of parties may be held liable for the costs of cleaning up sites contaminated by the actual or potential release of hazardous substances. Superfund liability is retroactive, strict, and (often) joint-and-several. Parties who are potentially liable include past and present owners and/or operators of the sites, the individuals or companies that generated the hazardous substances at the sites, those who arranged for the transportation of the hazardous substances to the sites, and those who arranged for disposal or treatment of the substances. Industries, government agencies, lending institutions (such as banks and savings-and-loans), small businesses, and even schools may fall into the category of liable parties as defined by the statute.

Congress gave EPA two tools for getting sites cleaned up: the first is its powerful enforcement authority to issue orders or bring legal action (which also results in bringing RPs to the table to negotiate with EPA); the second is the trust fund that EPA may use to finance cleanups itself (either at "orphan sites" where no RPs can be found or at other sites in order to initiate prompt cleanups for which costs will eventually be recouped from RPs).[11] Annual trust fund expenditures are approximately $1.6 billion (U.S. EPA 1993).

Option 2. Liability Release for Pre-1980 Wastes at Multiparty Sites (ASAP Pre-1980)

One of the most common complaints about Superfund concerns its retroactive imposition of strict liability. As examples of Superfund's unfairness, RPs often point to sites for which they are now liable but at which no hazardous substances have been deposited for many decades and where, they say, all waste handling and disposal was in compliance with the rules in place at the time. Although the courts have left little doubt about the constitutionality of retroactive liability, its application still rankles many RPs.

Under ASAP Pre-1980, liability would be waived at multiparty sites on the NPL for all RPs who sent wastes to these sites prior to the enactment of CERCLA in December 1980 (for convenience, we select a cutoff date of December 31, 1979, and refer to these as "pre-1980 wastes"). This liability waiver would apply to all future NPL sites as well. Under ASAP Pre-1980, RPs would continue to retain Superfund liability at all single-party sites and for waste disposed after December 31, 1979 at multiparty sites. RPs would not be released from liability if their waste disposal practices were illegal at the time of disposal.

Option 3. Proposed Clinton Administration Bill (H.R. 3800)

The major provisions of H.R. 3800 were designed to accomplish two goals: to reduce private-sector transaction costs (both for RPs and for insurers) and to contain liability for small businesses and local governments. In addition, the administration proposal injected a modicum of "fairness" into the current liability scheme by including factors in legislation to be used to allocate costs among RPs (U.S. EPA 1994). In order to estimate the costs of this proposal to the trust fund and on key sectors of the economy, we have simplified the provisions of the bill somewhat.[12]

Under H.R. 3800, the trust fund would pay for the shares of identifiable but insolvent parties—the orphan shares—at multiparty sites. With no way to estimate the financial implications of H.R. 3800's 10% cap on liability for generators and transporters of municipal solid waste, or the effect on responsible party commitments of relying on a set of factors to allocate liability among RPs, we do not estimate the financial implications of these latter two provisions separately, but instead include them in our overall estimate of the orphan share. Under H.R. 3800, the EIRF would have paid for 40% of cleanup costs borne by RPs for contamination resulting from hazardous substances disposed before 1986.[13] There would have been no reimbursement for the costs of contamination for disposal of hazardous substances that took place in 1986 or later.

ALTERNATIVE TAXES[14]

Three major taxes currently provide trust fund revenues: taxes on crude petroleum, taxes on chemicals, and taxes on corporations' alternative minimum taxable income (AMTI) in excess of $2 million, referred to as the corporate environmental income tax (EIT). The petroleum and chemicals that are subject to tax are not final products for sale to consumers, but rather are intermediate goods for use in producing some other final product. These taxes apply to the sale or use of these intermediate inputs; that is, whether they are purchased from another firm or produced within the same firm. In addition, these taxes are levied on specific goods that contain hazardous substances. The chemical taxes are levied on forty-two organic and inorganic chemical feedstocks, and on sixty-eight imported chemical substances (to the extent they were produced overseas using chemical feedstocks that would have been taxed in the United States). The imposition of the EIT, which was added in 1986 when CERCLA was amended, reflects the concern that hazardous substances are present in a broad range of goods and are used by firms other than those in the chemical and petroleum sectors. The EIT was imposed in 1986 as a major source of additional trust fund revenue.

Table 3 presents tax liabilities from each of the three tax categories for 1987 to 1991. The remainder of the trust fund moneys are from addi-

Table 3. Superfund Tax Liabilities, 1987–1991

	1987	1988	Tax ($ million) 1989	1990	1991	Total
Petroleum tax[a]	528.5 (45.8%)	547.6 (41.2%)	570.1 (43.2%)	545.2 (40.1%)	550.1 (41.4%)	2,741.5 (42.2%)
Chemical feedstock tax	273.3 (23.7%)	294.3 (22.1%)	277.3 (21.0%)	295.9 (21.7%)	299.6 (22.5%)	1,440.4 (22.2%)
Corporate environmental income tax (EIT)	351.3 (30.5%)	487.9 (36.7%)	471.8 (35.8%)	520.2 (38.2%)	479.3 (36.1%)	2,310.5 (35.6%)
Totals	1,153.1 (100.0%)	1,329.8 (100.0%)	1,319.2 (100.0%)	1,361.3 (100.0%)	1,329.0 (100.0%)	6,492.4 (100.0%)

[a]Excludes petroleum tax for the Oil Spill Liability Trust Funds.

Sources: Petroleum tax: U.S. Treasury Department 1994a; chemical feedstock tax: Dougherty and Gilson 1994; corporate environmental income tax: U.S. Treasury Department 1994b

tional allocations of general revenues, interest on the fund, and cost recovery actions. In 1991, the most recent year for which there is information on all three taxes, 41.4% or $550.1 million of trust fund taxes came from the petroleum tax, 36.1%, or $479.3 million came from the EIT, and almost 23.5% or $299.6 million came from the chemical feedstock tax. Thus, over 60% of the Superfund taxes came from taxes on two industry sectors—the chemical and petroleum industries.

We examine the implications of five different taxes as mechanisms for raising the increased trust fund revenues that would be required to finance changes in liability under Options 2 and 3. They include expansion of one or more of the three existing taxes under Superfund, the creation of a new excise tax on commercial property-casualty insurers, and the implementation of a value-added tax (VAT). Most proposals for raising additional trust fund revenues call for an increase in the EIT, a new tax on insurers, or some combination of the two: these are the other two mechanisms we examine below. We look at the implications of using a small increase in a VAT, should that tax be adopted for some other purpose, such as financing the Clinton administration's health care proposal.

To help determine the likely incidence of the current Superfund taxes, we refine and use an input-output model with more than 500 industrial and commercial sectors. This model is used to determine the incidence of the current taxes as well as any increase in the current taxes or new taxes to finance a revised liability scheme. We evaluate the effect of raising an additional $1 billion annually from each of the tax mechanisms examined.

The model takes into account the indirect effects of each taxing mechanism examined. For example, a tax on one input may differentially affect the prices of different outputs. In most cases, chemical feedstocks and other affected commodities are not purchased directly by consumers, but are purchased by other firms for use in their production processes. In order to assess who bears the ultimate burden of these initial costs, we determine which final consumption goods are affected. The input-output model is used to estimate price changes sector by sector, taking into account these indirect effects. If only certain industries use chemical feedstocks, for example, then an additional tax on chemicals may be found to affect some prices more than others. On the other hand, if most industries use petroleum, or electricity generated from petroleum, then a tax on petroleum may have more diffuse effects. What we find, however, is that the effects of generating $1 billion from any of these taxes are quite diffuse, with little major effect on any particular industry sector. We also find, however, that the administrative and compliance costs for some of these taxes—including the existing three Superfund taxes—may be quite large in comparison to the revenues generated.

WHO PAYS UNDER ALTERNATIVE LIABILITY SCHEMES?

For each liability alternative, we look at three major effects: the distribution of cleanup costs between RPs and the trust fund; the distribution of cleanup costs among key industries; and, finally, the impact of each alternative on total transaction costs and how these costs are distributed among industries. The estimates of cleanup costs do not include the day-to-day costs of operating the Superfund program, which amount to approximately $1 billion annually.

Our analysis focuses primarily on the annual cost of each liability alternative, rather than the cumulative costs. This is because of the need to compare the tax implications to the costs of a liability-based system. We arrive at annual costs quite simply: by dividing by ten the remaining cleanup costs and estimated transaction costs for the current NPL. Averaging the remaining costs of cleaning up the current NPL over the next ten years assumes that the minimum site cleanup duration is ten years, the maximum, twenty-four years.

Distribution of Cleanup Costs between RPs and the Trust Fund

Under the current Superfund program (Status Quo), the cost of cleaning up all nonfederal facility NPL sites is $21.4 billion, or $2.1 billion annually. Under Status Quo, 27% of cleanup costs, $585.0 million annually, are paid for by the trust fund. These funds pay for 100% of the cleanup costs for eighty-four orphan sites—that is, sites where there are no viable RPs. The trust fund also covers just under 25% of the costs of cleanup at all other sites. While RPs often take the lead for site cleanups, in most cases they do not cover 100% of cleanup costs.

As is shown in Figure 3, needed trust fund revenues increase quite dramatically for both of the liability alternatives we examine, as compared with Status Quo. If retroactive liability is eliminated for legal disposal at multiparty sites for hazardous substances disposed of before 1980, as in ASAP Pre-1980, the trust fund would need to generate revenues to pay for 70% of cleanup costs, or $16.4 billion over ten years. This increase in government-implemented cleanups results in a $2.2 billion increase in the total cleanup bill under ASAP Pre-1980, as compared with Status Quo, because of the relative inefficiency of government cleanups as compared to those implemented by RPs. Shifting the date for elimination of retroactive liability at multiparty sites to 1987 would mean that an additional $3.4 billion in cleanup costs are paid for by the trust fund, rather than by RPs at sites, as compared to Status Quo.

Assessing who pays under H.R. 3800 is a two-step process. First, we estimate the effect of changes in liability on the trust fund. Second, we esti-

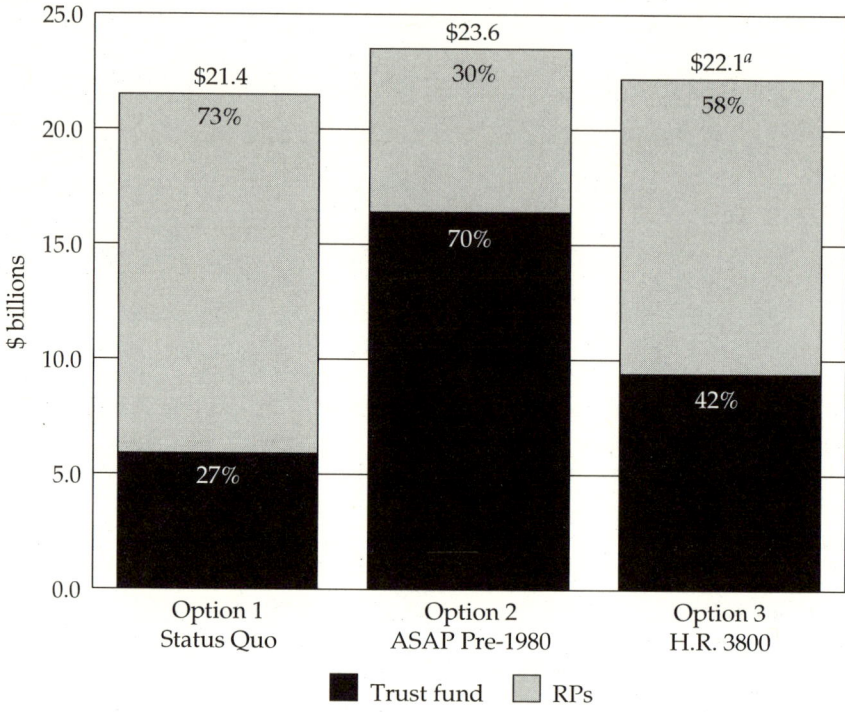

Figure 3. Estimated Remaining Cleanup Costs: RPs versus Trust Fund
[a]Costs to the EIRF are included in RP cleanup costs in Option 3.
Source: RFF NPL Database 1994.

mate the effect on RPs of reimbursement from the EIRF. As compared to Status Quo, the trust fund share of total remaining cleanup costs increases by $3.5 billion to a total of 42% of cleanup costs, to pay for orphan shares. This increase, 13.6% of the cleanup costs at multiparty sites, is to cover the shares that would otherwise be allocated to identifiable but insolvent parties at these sites.[15] Actual trust fund responsibility could be even larger under H.R. 3800, if, in fact, the 10% cap on generators and transporters of municipal solid waste results in additional orphan shares.

The increased trust fund payments leave RPs with 58% of the costs of cleanup for the current NPL under H.R. 3800. However, as shown in Table 4, under this option, $367.5 million (17%) of remaining cleanup costs are reimbursed by the EIRF, leaving RPs with a cleanup bill of $911.1 million annually, or 41% of total cleanup costs.[16]

None of the estimates presented so far includes payments for natural resource damage—which could amount to billions of dollars—or payments for any sites added to the NPL in the future. Nor do these

Table 4. Estimated Annual Cleanup Costs for Each Option

Estimated financing mechanism	Annual cleanup costs ($ millions)		
	Option 1. Status Quo	Option 2. ASAP Pre-1980	Option 3. H.R. 3800
Liability	1,559.1 (73%)	711.8 (30%)	911.1 (41%)
Trust fund	585.0 (27%)	1,644.0 (70%)	935.5 (42%)
EIRF	–	–	367.5 (17%)
Total annual cleanup expenditures	2,144.1 (100%)	2,355.8 (100%)	2,214.1 (100%)

Source: RFF NPL Database 1994.

estimates include the costs of compensating RPs or their insurers past costs or the day-to-day costs of implementing the Superfund program. Finally, both ASAP Pre-1980 and H.R. 3800 include major new initiatives to improve cleanup standards and increase community involvement at sites. All of these efforts cost money. Thus, the preceding estimates of the total cost of these options to the trust fund represent a lower bound. In Table 5 we include rough estimates of these other Superfund costs, in addition to the foregoing estimates of cleanup costs. Based on the assumptions noted in Table 5, total program costs (not including transaction costs borne by RPs or insurers) are $3.7 billion for Status Quo, and rise to just over $4.5 billion annually for both ASAP Pre-1980 and H.R. 3800.

The largest single component of these other costs is the $1 billion a year under Status Quo to pay for the day-to-day costs to the government of implementing the Superfund program, costs that we increase by $150 million a year under ASAP Pre-1980 and H.R. 3800 to account for new program initiatives noted above. The increased burden of these additional costs is largest for the trust fund, which generates $1.6 billion annually under Status Quo, but must cover $3.6 billion in annual costs under ASAP Pre-1980. Under H.R. 3800, the trust fund would need to generate revenues of $2.2 billion annually, according to our estimates, and the EIRF would need to provide $1.2 billion in revenues each year.

Distribution of Cleanup Costs among Major Industry Sectors

Under Status Quo, the chemical and allied products industry bears the largest percentage of cleanup costs as a result of Superfund liability of any industry, 25% of the $1.6 billion in annual cleanup costs borne by RPs, or $394.4 million. This is not surprising since 8% of the nonfederal NPL sites are chemical manufacturing sites, where the estimated average site cleanup cost is $41.1 million. The next largest share, 11%, or $174.5 million each year, is borne by the mining industry. Recycling facilities, such as battery and oil recyclers, account for another 10% of the cleanup costs borne by

Table 5. Estimated Annual Total Program Costs for Each Option, Excluding RP and Insurer Transaction Costs

	Annual total program costs ($ millions)		
Type of costs	Option 1. Status Quo	Option 2. ASAP Pre-1980	Option 3. H.R. 3800
RPs			
Site study/cleanup costs[a]	1,715.0	783.0	1,002.2
Adder for NRD costs[b]	343.0	156.6	200.4
Subtotal	2,058.0 (56%)	939.6 (21%)	1,202.6 (29%)
Trust fund			
Site study/cleanup costs[a]	643.5	1,808.4	1,029.1
Adder for NRD and other costs[c]	0.0	400.0	0.0
Program management[d]	1,000.0	1,150.0	1,150.0
Payment to RPs of past costs[e]	0.0	250.0	0.0
Subtotal	1,643.5 (44%)	3,608.4 (79%)	2,179.1 (53%)
EIRF			
Site study/cleanup costs[a]	–	–	404.3
Adder for NRD and other costs[c]	–	–	175.0
Payment to RPs of past costs[e]	–	–	175.0
Subtotal	–	–	754.3 (18%)
Total program costs[f]	3,701.5 (100%)	4,548.0 (100%)	4,136.0 (100%)

[a] We add 10% to our estimate of annual site study/cleanup costs for the current NPL to account for future NPL sites.

[b] We assume RPs' NRD costs to be 20% of RP site study and cleanup costs under each option.

[c] This adder is a rough estimate of the costs of natural resource damages, non–NPL removals, and defense costs. There is little basis to estimate likely NRD costs, the largest component of the costs on this line. The cost of natural resource damages could be substantially higher than is assumed here.

[d] Both the ASAP and H.R. 3800 proposals include many new program initiatives, which we account for by adding $150 million to the $1 billion annual program management total.

[e] We assume that by the time Superfund is reauthorized, RPs will have spent approximately $5 billion on site study and cleanup activities. For each alternative, a different percentage of these past costs would be reimbursed by the trust fund and EIRF.

[f] Does not include RP or insurer transaction costs.

Source: RFF NPL Database 1994.

RPs as a result of the current liability scheme. After that, as shown in Table 6, costs are distributed among several other key industrial sectors.

Under all three liability options, the chemical and allied products industry bears the largest share of cleanup costs resulting from Superfund liability of any industry. The relative percentage of total cleanup costs borne by each industry sector remains almost constant under each liability option. Not surprisingly, all sectors see a decrease in their annual cleanup costs under each alternative liability scheme when

Table 6. Estimated Annual RP Cleanup Costs by Sector for Each Option

Industry categories	Annual cleanup costs ($ millions)		
	Option 1. Status Quo	Option 2. ASAP Pre-1980	Option 3. H.R. 3800
Mining[a]	174.5 (11%)	73.2 (10%)	103.6 (11%)
Lumber and wood products, except furniture[a]	98.1 (6%)	44.2 (6%)	57.6 (6%)
Chemicals and allied products[a]	394.4 (25%)	187.7 (26%)	229.5 (25%)
Petroleum refining and related industries[a]	75.3 (5%)	30.7 (4%)	43.2 (5%)
Primary metal industries[a]	118.0 (8%)	55.7 (8%)	71.6 (8%)
Fabricated metal products, except machinery and transportation equipment[a]	79.7 (5%)	42.4 (6%)	48.3 (5%)
Electronic and other electrical equipment and components, except computer equipment[a]	57.7 (4%)	32.2 (5%)	34.6 (4%)
All other manufacturing[a]	94.7 (6%)	45.0 (6%)	55.0 (6%)
Miscellaneous[b]	91.4 (6%)	45.9 (6%)	54.2 (6%)
Recycling[c]	157.9 (10%)	69.3 (10%)	87.9 (10%)
Not attributed	217.4 (14%)	85.5 (12%)	125.5 (14%)
Annual RP cleanup cost[d]	1,559.1 (100%)	711.8 (100%)	911.1 (100%)

[a]Industry sectors are based on U.S. OMB 1987.
[b]Miscellaneous includes all Standard Industrial Classification (SIC) codes not specifically noted above, such as transportation, services, and public administration.
[c]Category developed by RFF to classify activities not captured by the SIC.
[d]Numbers may not add due to rounding.

Source: RFF NPL Database 1994.

compared to Status Quo as the percentage of total cleanup costs borne by RPs decreases.

The most dramatic reduction of cleanup costs resulting from Superfund liability occurs under ASAP Pre-1980. For example, the cleanup costs borne by the chemical and allied products industry decrease from a high of almost $400 million annually under Status Quo to less than $190 million under ASAP Pre-1980. Of course, these decreased cleanup costs must be examined in comparison to any increased taxes that would be paid by the chemical industry to finance an increased trust fund. Under H.R. 3800, the chemical and allied products industry costs are reduced as well, to around $230 million. This pattern is repeated for all industry sectors.

Looking at the distribution of liability-induced cleanup costs on an annual basis, it appears that the total burden on each sector is not that large. Even for chemical and allied products, the annual cost of site cleanups is estimated to be under $500 million dollars, not including transaction costs. Of course, smaller burdens may have a larger effect on smaller, less well-financed industries, such as mining and fabricated metal products. Perhaps most surprising is the relatively large percentage of costs attributed to the recycling industry. This industry is not included in the Standard Industrial Classification, but rather is a category we developed to group those facilities whose primary service is recycling batteries, waste oil, drums, and so forth. Under Status Quo, these facilities pay 10% of cleanup costs attributed to RPs. The percentage share borne by recyclers stays the same under Options 2 and 3, although the actual financial exposure is cut almost in half under H.R. 3800—to $87.9 million on an annual basis—as compared to Status Quo, and falls to a low of $69.3 million a year under ASAP Pre-1980.

Total Private-Sector Transaction Costs and Their Distribution among Industries

Figure 4 shows estimated RP transaction costs as a percentage of total trust fund and responsible party site costs. Under the current liability scheme, we estimate total transaction costs to be $418.5 million, or 16% of total costs of $2.6 billion annually. This estimate is lower than that presented by RAND because RP transaction costs are computed as a share of total—that is, fund and RP—cleanup costs and RP transaction costs. Our estimate of RP transaction-cost share as a percentage of *total* RP costs—which is the way that RAND presents transaction-cost share—is 21%, which is in line with RAND's estimates. Transaction costs fall under ASAP Pre-1980 to 7% of total costs. RP transaction costs under H.R. 3800, while lower than under Status Quo at 13% of total

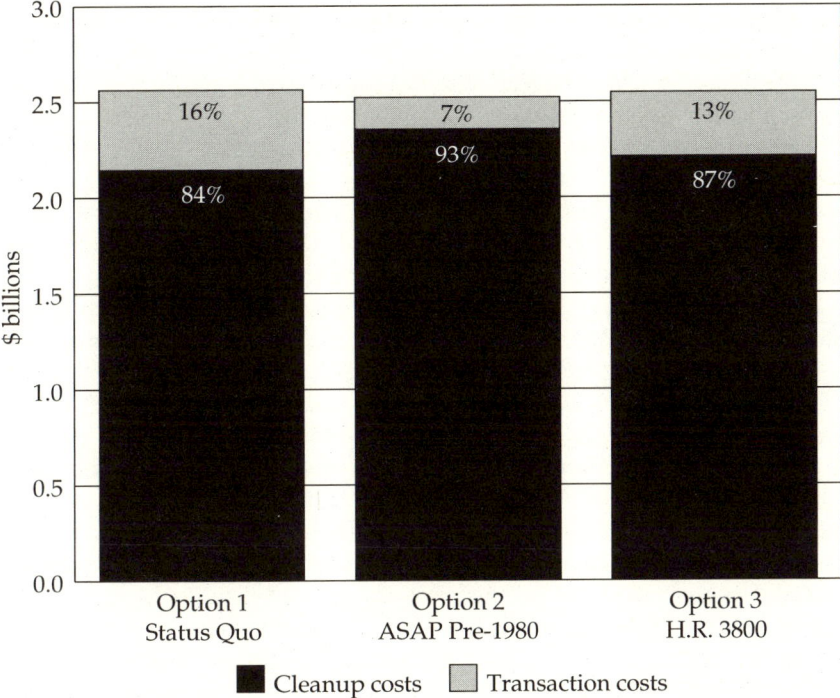

Figure 4. Estimated Annual RP Transaction Costs as a Percentage of Total Costs
Source: RFF NPL Database 1994.

costs, are much higher than under ASAP Pre-1980, which completely eliminates liability at many sites.

Not surprisingly, the distribution of private-sector transaction costs mirrors the distribution of cleanup costs among industry categories. The relative distribution of transaction costs among sectors changes little under each option, but there are sizable reductions in the actual dollars spent on transaction costs. For example, transaction costs for the chemical and allied products industry would be reduced from $97.6 million under Status Quo to $38.5 million under ASAP Pre-1980.

Which Industries Pay the Most for Superfund?

Much of the impetus for reforming Superfund liability is based on the premise that cleanup and transaction costs are just too high. Our analysis suggests that, in fact, the annual cost of Superfund is relatively small—with annual expenditures from both the trust fund and RPs, including

RP transaction costs—just over $2.5 billion. The largest costs are borne by the chemical industry, which has the largest direct costs under the current liability scheme ($492.0 million a year) and pays just over 22% of the Superfund taxes: $299.6 million annually. The petroleum industry, though seldom an RP, pays the next largest share of Superfund costs, as the petroleum tax generates over $550 million each year.

Other industry sectors pay a much smaller share of total costs, although these expenses may still be overwhelming for some industries, such as mining and wood preserving, to shoulder. Still, barring a dramatic increase in the expected costs of site cleanup or the total number of sites cleaned up under the Superfund program, annual cleanup and transaction costs borne by RPs as a result of Superfund's liability standards under Status Quo amount to just under $2 billion for the current NPL. While it is likely that new sites will be added to the NPL in the future, the costs of cleaning up these new sites will likely not be incurred for some years. The largest unknown is the financial cost of natural resource damage, which many fear will turn out to be a sleeping giant.

The impact on property-casualty insurers is more difficult to estimate. The insurance sector pays a very small amount of the EIT, less than 3%. Insurers have spent little to date on cleanup costs and much more on legal fees, both to cover their own litigation with their insureds and to reimburse their insureds' legal defense costs. Of course, to the extent that courts do require insurers to cover Superfund-related claims, the costs to RPs decrease. It is still too soon to know the likely financial impact of Superfund on the insurance sector. If, for example, insurers are required to pay only 5% of total cleanup costs for the current NPL of $33 billion, then their total cleanup responsibility would be under $2 billion, spread out over some number of years. On the other hand, if they are found responsible for 50% of cleanup costs, their financial responsibility would be $16.5 billion, not including transaction costs. And, to the extent that insurers are successful in repudiating the claims of their insureds, the RPs, a larger percentage of their costs will be transactional in nature.

Almost all the proposals for reforming Superfund liability call for new taxes to generate needed additional revenues, either for an increased trust fund or for some kind of insurance resolution fund, such as the EIRF. Our analysis suggests that none of the tax options under consideration will have large or negative effects on the economy as a whole, in large part because the amount of revenue to be raised is relatively insignificant in terms of the U.S. economy as a whole. Of concern, however, are the administrative and compliance costs that will be incurred for any new tax levied. All taxes have their own form of "transaction" (that is nonproductive) costs, the costs of administering and enforcing the tax. Thus, proposals that call for raising $100 million from one tax, and a few hundred million dollars from another, are not good tax policy. These costs must be weighed against the revenues generated in determining their cost-effectiveness as compared to the current liability approach. From an efficiency standpoint, it would make more sense to generate any additional fund revenues from one of the existing Superfund taxes than to implement a new tax.

What is inescapable is that what motivates individual companies to seek Superfund liability reform is not the cost to any industry segment, but rather is the extent of their individual future exposure. Some insurers have far greater financial exposure than others, just as some RPs are involved at extremely expensive sites, while others are not. The company-specific nature of the Superfund "problem" makes it difficult to examine and difficult to figure out equitable solutions.

Although there is heated debate about whether Superfund's liability scheme should be changed and, if so, how, the ferociousness of the debate would appear to be grossly disproportionate to the economic importance of the problem. Most experts estimate that at most the Superfund program is costing the national economy $5 billion annually, not including the cost of cleaning up federal facilities. More recent estimates by EPA and others—including RFF—put the estimates of annual cleanup expenditures for NPL sites much lower, at around $3 billion. Even if one adds $1 billion for program operations and throws in another $1 billion for private-sector transaction costs (certainly on the high side), the total annual cost of the Superfund program is still $5 billion. This is less than 10% of estimated current annual compliance spending under either the Clean Air Act or the Clean Water Act, and is probably less than the annual compliance expenditures necessitated by the Resource Conservation and Recovery Act, the Federal Insecticide, Fungicide, and Rodenticide Act, or the Safe Drinking Water Act.

PUTTING SUPERFUND IN CONTEXT

In reality, the costs associated with Superfund are probably not that large. In terms of the national economy they show up not at all. Of course, for some industries and some companies—both large and small—the costs of Superfund are potentially devastating, and this possibility is one reason why the program attracts a lot of attention. The real focus of the Superfund debate, though, should be on what we as a society are getting for the $3 billion spent annually on site studies and cleanup. Whether this money is well spent is a far more important policy question than whether we are losing too much in transaction costs.

ACKNOWLEDGMENTS

This chapter was written with the help of Karen S. Terry, a research assistant at Resources for the Future, and is based on research done with some of my colleagues: Paul Portney, Vice President and Senior Fellow at Resources for the Future; Robert Litan, U.S. Department of Justice (on leave from the Brookings Institution); and Don Fullerton, University of Texas–Austin. The full discussion and presentation of our research was published in *Footing the Bill for Superfund Cleanups: Who Pays and How?*, published jointly by RFF and the Brookings Institution in January 1995. This work has been funded in part by grants from the U.S. EPA's Office of Policy, Planning, and Evaluation and Office of Solid Waste and Emergency Response (grants CR815934 and CR820740).

ENDNOTES

[1] Some portion of the 230 sites noted earlier are also assigned to the "not attributed" category.

[2] This research found that DOE projects cost an average of 32% more, take longer to complete, and have larger cost overruns than comparable projects implemented by private industry (DOE 1993).

[3] The Clinton administration assumed a 20% cost-savings for RP-implemented cleanup of nonfederal facility sites in its analysis of the cost of the administration bill (Glauthier 1994).

[4] The average costs of site studies ($4.2 million) and cleanup actions ($22.0 million) for NPL sites are in "as built" dollars; average operation and maintenance (O&M) costs of $2.9 million per site are also included. The O&M costs are the net present value of future O&M activities at the site.

[5] The number of RPs is defined here as the total number of parties who are potentially liable at the site, not as the number that have been held financially liable by the government.

[6] In this estimate is included the money EPA has spent on remedial investigation/feasibility studies, remedial designs, and remedial actions (Evans 1994).

[7] This chapter discusses only the aspects of both proposals that affect Superfund financing; that is, the liability and funding schemes. In this chapter, the proposal based on the initial bill first introduced in February 2, 1994 are described; there have been subsequent amendments that are not addressed.

[8] *Municipal solid waste* refers to the category of waste generated, not to the status—public or private—of the waste generator.

[9] Other groups have proposed modifications to the administration's EIRF proposal (U.S. House 1994).

[10] These three alternatives and two other liability alternatives are evaluated in Probst and others 1995.

[11] Throughout this chapter, all government action, including enforcement, is referred to as stemming from EPA, although in many cases the U.S. Department of Justice and state agencies are also involved.

[12] For example, the cost of changes in liability for municipal solid waste are not estimated separately

but, rather, these costs are assumed to be part of the "orphan share."

[13] Because it is unlikely that all RPs will participate in the EIRF, we assume 40% reimbursement for 85% of all RP costs, rather than of 100% of these costs.

[14] This section was written by Professor Don Fullerton of Carnegie Mellon University. This chapter presents only a brief summary of our analysis of the impact of the alternative tax mechanism. A much fuller discussion is included in Probst and others 1995.

[15] This percentage is based on an analysis conducted by EPA.

[16] It is assumed here that 85% of RPs will agree to settle with the EIRF and that on average the EIRF reimburses RPs for 40% of cleanup costs.

REFERENCES

Acton, Jan Paul, and Lloyd S. Dixon. 1992. *Superfund and Transaction Costs: The Experiences of Insurers and Very Large Industrial Firms*. Santa Monica: RAND.

Colglazier, E.W., T. Cox, and K. Davis. 1991. *Estimating Resource Requirements for NPL Sites*. Knoxville, TN: University of Tennessee, Waste Management Research and Education Institute.

Dougherty, Charlotte, and Elizabeth Gilson 1994. *Economic Impacts of Superfund Taxes*. Prepared for the Office of Policy Analysis, U.S. Environmental Protection Agency, February, 1994. Cambridge, Mass.: Industrial Economics, Inc.

Evans, David S. 1994. Letter to Katherine N. Probst. Washington, D.C., March 31, 1994.

Glauthier, T.J. 1994. Memorandum to Katie McGinty, Bo Cutter, and Robert Sussman. Washington, D.C., February 2, 1994.

Probst, Katherine N., and Paul R. Portney. 1991. *Assigning Liability for Superfund Cleanups: An Analysis of Policy Options*. Washington, D.C.: Resources for the Future.

Probst, Katherine N., Don Fullerton, Robert Litan, and Paul R. Portney. 1995. *Footing the Bill for Superfund Cleanups: Who Pays and How?* Washington, D.C.: The Brookings Institution and Resources for the Future.

RFF NPL Database. 1994. Washington, D.C.: Resources for the Future.

Russell, Milton, E. William Colglazier, and Mary R. English. 1991. *Hazardous Waste Remediation: The Task Ahead*. Knoxville: University of Tennessee, Waste Management and Research and Education Institute.

U.S. DOE (Department of Energy). Office of Environmental Restoration and Waste Management. 1993. *Project Performance Study.* Reston, Virginia: Independent Project Analysis, Inc.

U.S. EPA (Environmental Protection Agency). 1993. *Superfund Management Report.* December 6, 1993. Washington, D.C.: U.S. EPA.

———. 1994. *Summary of the Majority of the Provisions in the Administration's Proposal to Reform Superfund.* Washington, D.C.: U.S. EPA.

U.S. House. 1994. Committee on Energy and Commerce. Subcommittee on Transportation and Hazardous Materials. *Legislative Hearing to Discuss Title VIII of H.R. 3800, Environmental Insurance Resolution Fund. Statement of Edward Pollack.* 103rd Cong., 1st sess., 17 March, Washington, D.C.: U.S. Government Printing Office.

U.S. OMB (Office of Management and Budget). 1987. *Standard Industrial Classification Manual.* Springfield, Virginia: National Technical Information Service.

U.S. Treasury Department. 1994a. *Environmental Tax Statistics and Tabulations, August 1992 and 1993.* Washington, D.C.: U.S. Treasury Department, Statistics of Income.

———. 1994b. *Corporate Income Tax Returns (Publication 16): Returns of Active Corporations.* Washington, D.C.: U.S. Treasury Department, Statistics of Income.

PART IV:
Transaction Costs

7
The Transaction Costs Generated by Superfund's Liability Approach

Lloyd S. Dixon

When Congress passed Superfund in 1980 to clean up the nation's worst inactive hazardous waste sites, it adopted a liability-based approach. Rather than fund cleanups entirely with tax revenues, Congress held as liable for cleanup the parties that generated or transported the hazardous materials at a site or that owned or operated the site.

The liability approach has strong appeal to many interest groups and policymakers. First, it attempts to make the polluter pay for cleanup. Second, it does not require a large increase in tax revenues to fund a traditional public works program. Finally, it creates strong incentives for firms to more carefully handle hazardous substances, although such incentives require only prospective, as opposed to retroactive, liability.

Many participants in the process and other stakeholders are concerned, however, that the advantages of a liability-based approach come at the cost of high transaction costs (see, for example, Hedeman, Cannon, and Friedland 1991). In this chapter, I discuss what several studies at RAND have revealed about the transaction costs generated by the Superfund process. I start by defining transaction costs and then discuss why Superfund might be expected to generate substantial transaction costs. In the next section, I present evidence from three RAND studies on the size of these transaction costs. I then compare Superfund transaction costs with those generated in tort litigation and finally close with a brief conclusion.

WHAT ARE TRANSACTION COSTS?

Transaction costs, unlike the costs to investigate and remediate the site, do not contribute directly to the cleanup process; instead, they are incurred in the process of assigning liability among the various parties involved at a site. Separating expenditures into cleanup costs and transaction costs, however, is sometimes difficult. Legal costs are generally transaction costs, but nonlegal costs can be either transactional or cleanup in nature. For example, engineering studies done by potentially responsible parties (PRPs) to characterize the wastes at a site are transactional if their purpose is to assist in the search for other PRPs or to contest a remedy chosen by the U.S. Environmental Protection Agency (EPA). In contrast, engineering studies are not transactional if they contribute to a better understanding of how to clean up the site.

In the three RAND studies on transaction costs (Acton and Dixon 1992; Dixon, Drezner, and Hammitt 1993; Dixon 1994), we classified nonlegal PRP expenditures as transaction costs if they did not have some sort of government approval, such as a consent decree or an administrative order. Activities done without an agreement are likely to duplicate government efforts. In such situations, one of the two activities is not necessary to the cleanup process, and we arbitrarily classified the PRP expenditures as transaction costs.

Some legal costs related to Superfund sites are excluded from our analyses. We excluded expenditures related to lawsuits for bodily injury or property damage brought by residents or landowners living near Superfund sites. Although the costs related to these lawsuits are potentially substantial, they are not related directly to site cleanup and could have occurred under federal and state tort law in the absence of Superfund. Consequently, we do not treat them as resulting from Superfund's liability approach.

INTERACTIONS AMONG KEY PLAYERS GENERATE TRANSACTION COSTS

The cleanup process generates transaction costs because the law, and the way the EPA implements it, creates a complex set of often contentious interactions among the many different players. Figure 1 illustrates the key Superfund players and the major interactions. The five major sets of interactions are described below.

PRP-Government Interactions

The Superfund process requires the government and the PRPs to interact in a variety of ways. At each site on the National Priorities List (NPL), EPA attempts to negotiate cleanup agreements with or to recover cleanup costs from the PRPs at the site. Transaction costs are

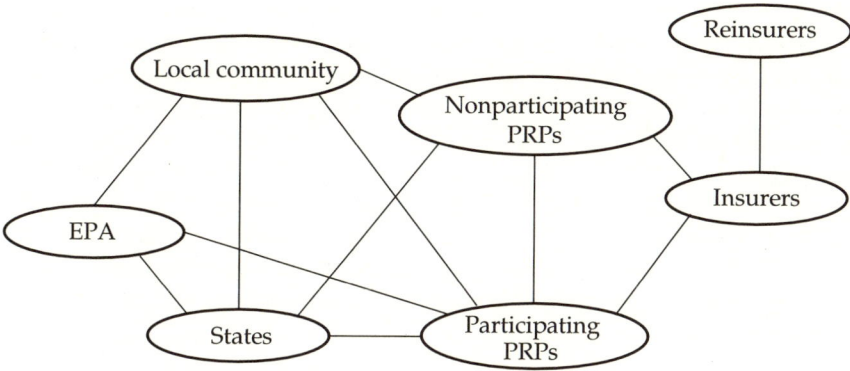

Figure 1. Players and Interactions in the Superfund Process

generated in arguing over the cleanup standards and remedy; in negotiating settlements; and, when negotiations fail, in enforcement actions. States frequently are involved in setting cleanup standards and selecting remedies, further complicating the process.

PRP-PRP Interactions

Negotiations take place among the participating PRPs, who must allocate liability among themselves. Because EPA wants to negotiate with a PRP committee rather than with individual PRPs, PRPs must agree on common negotiating positions. In addition, some PRPs attempt to avoid liability at the site, and the participating PRPs seek to recover costs from these nonparticipating PRPs, usually through the courts.

PRP-Insurer Interactions

PRPs also negotiate with their insurers, to whom they turn for reimbursement of legal and cleanup costs. Such claims typically are brought under comprehensive general liability and commercial multiperil policies, but whether or not the policies cover these claims is usually hotly contested.

Insurer-Reinsurer Interactions

Insurers may seek compensation for their costs from their reinsurers. Even though many insurers have notified their reinsurers of potential claims, observers of the reinsurance market assert that reinsurer expenditures have begun to grow only recently—perhaps because of the very high deductible for most reinsurance policies. However, conflict between insurers and reinsurers may increase over time if insurer costs rise.

Community-PRP and Community-Government Interactions. The local community around a site interacts with both the federal and state governments and the PRPs. (For the sake of simplicity, local government is considered part of the local community.) The community may want to have input into remedy selection and cleanup standards. It may also oppose cleanup decisions that have been made or slow or stop the cleanup process.

HOW LARGE ARE TRANSACTION COSTS?

How large are the transaction costs generated by this complex web of often contentious interactions? Three RAND studies suggest that they are substantial. In this section, I first discuss evidence on the size of PRP and insurer transaction costs through 1991. Then I present estimates of combined PRP and insurer transaction costs through 1991. I then discuss what transaction costs ultimately might be when cleanup is complete at all NPL sites and finally discuss government transaction costs.

PRP Transaction Costs

A 1992 RAND study of five very large industrial firms—those with annual revenues over $20 billion—found that transaction costs were 19% of their total expenditures at forty-nine NPL sites between 1984 and 1989 (Acton and Dixon 1992, 48). A 1993 RAND study of 108 smaller PRPs at eighteen NPL sites found that transaction costs were 19% of total expenditures between 1981 and 1991 for firms with annual revenues between $1 billion and $20 billion and 15% for firms with annual revenues between $100 million and $1 billion (Dixon, Drezner, and Hammitt 1993, 23).[1] Transaction-cost shares were much higher for smaller firms. (*Transaction-cost share* is the ratio of transaction to the sum of transaction costs and cleanup costs.) Those with annual revenues between $15 million and $100 million and those with revenues less than $15 million both averaged shares of 60% (see Table 1).

In the 1993 RAND study, we developed a model of firm expenditures to estimate overall PRP expenditures and transaction costs at the eighteen study sites. Because a sizable number of the firms in our sample had very small expenditures and relatively few firms had very large expenditures, we developed two separate two-part models: one for transaction costs

Table 1. PRP Transaction-Cost Share by Firm Size

1991 annual revenues	1992 study (percent, 1984–1989)	1993 study (percent, 1981–1991)
>$20 billion	19	–
$1–20 billion	–	19
$100–1,000 million	–	15
$15–100 million	–	60
<$15 million	–	60

Sources: 1992 study: Acton and Dixon 1992, 48; 1993 study: Dixon, Drezner, and Hammitt 1993, 30.

and another for cleanup costs.[2] Each two-part model first modeled the firm's decision to make any expenditures at a site during a specified cleanup phase as a function of site and firm characteristics and then modeled the natural logarithm of expenditures given positive expenditures as a function of the same site and firm characteristics.[3]

Based on these models, we estimated that transaction costs accounted for 32% of total private-sector PRP expenditures at the eighteen study sites through 1991. As might be expected given that we collected data from relatively few of the 3,650 PRPs at the eighteen study sites, there was considerable uncertainty in this estimate. The 90% confidence interval around the point estimate ranged from 20% to 44% (Dixon, Drezner, and Hammitt 1993, 46).

How representative is the estimated transaction-cost share at the eighteen study sites of private-sector transaction-cost share at all NPL sites through 1991? Since the eighteen study sites were chosen randomly from a set of sites where we had reason to believe there had been nontrivial private-sector expenditures through 1991, 32% may well be a good estimate of the transaction-cost share for total PRP expenditures through 1991.[4] However, eighteen is not a large number of sites, and additional work is warranted to more formally project the findings from these few sites to the entire NPL.

Impact of Site and Firm Characteristics on Transaction Costs

In both the 1992 and 1993 RAND studies, we found that transaction costs and transaction-cost share vary importantly with site and firm characteristics. An important factor in explaining any given firm's expenditures and transaction-cost share at a particular site is its volumetric share of wastes. (A firm's *volumetric share* is the proportion of the waste by volume it contributed to the site.) Important factors in explaining the overall expenditures and transaction costs of *all* PRPs at a site are the number of PRPs at the site and the expected site cleanup cost. Each of these factors is considered in turn.

Firm Volumetric Share. We found a strong relationship between a firm's volumetric share at a site and both its total expenditures and transaction-cost share at a site: firms with higher volumetric shares spend more at a site but have lower transaction-cost shares. Volumetric share and not firm size appears to be the most important factor in explaining expenditures and transaction-cost share. The correlation between firm size and transaction-cost share apparent in Table 1 appears not to be due to firm size itself, but rather to the tendencies of smaller firms to account for a small share of the waste at a site and firms with small volumetric shares to have high transaction-cost shares (Dixon, Drezner, and Hammitt 1993, 41). Regression analysis reveals that the relationship between firm size and transaction-cost share disappears once volumetric share is taken into consideration.

The relationship between volumetric share and transaction-cost share suggests that certain fixed costs are attached to mounting a defense at a Superfund site. Furthermore, this relationship suggests that firms accounting for a small share of the waste at a site usually do not incur sufficiently large cleanup costs to offset these fixed costs to the same extent as firms with large volumetric shares.

The relationship between volumetric share and transaction-cost share also suggests that small-volume contributors at a site incur transaction costs that are high in proportion to the wastes they contribute to the site. This indeed appears to be the case at the eighteen study sites in our 1993 study. As shown in Table 2, we found that firms with volumetric shares less than 1% accounted for 95% of the 3,650 PRPs at the study sites and 56% of the transaction costs, but only 26% of the cleanup costs through 1991. These relationships suggest that efforts to expedite settlements for small-volume contributors may be particularly effective in reducing transaction costs.

Number of PRPs. We found that transaction-cost shares are lower at sites with fewer PRPs when other site characteristics are held constant (see Table 3).[5] This is to be expected since the transaction costs gener-

Table 2. Expenditures through 1991 of PRPs with Volumetric Shares of Less Than 1% at Eighteen NPL Sites

Type of expenditure	Percent[a]
Share of total PRPs	95
Transaction-cost share	54
Share of total site transaction costs	56
Share of total site cleanup costs	26

[a]Percentage of the 3,650 PRPs at the eighteen study sites
Source: Dixon 1994, Table 2.2.

Table 3. Transaction-Cost Share by Number of PRPs at the Site

	Transaction-cost share (percent)
1992 Study	
Single-PRP sites	5
Multiple-PRP sites	24
1993 Study	
≤15 PRPs	14
16–100 PRPs	36
>100 PRPs	34

Source: Dixon 1994, Table 2.1.

ated by PRP-PRP interactions are likely to be lower with fewer parties. Note, however, that even at sites with few PRPs, other important sources of transaction costs remain, such as those generated by disputes between PRPs and the government over remedy or between PRPs and insurers over insurance coverage.

Expected Site Cleanup Costs. We also found in the 1993 study that transaction costs are higher, but transaction-cost share is lower, at sites with higher expected cleanup costs when other factors are held constant. It seems reasonable that the higher stakes at more expensive sites induce higher transaction costs. The fact that cleanup costs rise even faster at more expensive sites, thus causing transaction-cost share to fall, again suggests that transaction costs have both a fixed component and a component that varies somewhat with site and firm characteristics.

Insurer Transaction Costs

The 1992 RAND study collected data from four national insurance carriers on claims involving inactive hazardous-waste sites. Both large and medium-sized insurance carriers were represented. In all, these insurers had received over 10,000 claims for cleanup at inactive hazardous-waste sites through 1989 (Acton and Dixon 1992, 16, 18). We estimated that insurers nationwide spent about $370 million on cleanup claims in 1989, about $150 million of which was at NPL sites.

The vast majority of insurer outlays through 1989 were transaction costs. Table 4 breaks down insurer outlays into indemnity payments and transaction costs. Indemnity payments are made to policyholders, generally for site cleanup. Insurer transaction costs have three main components: the costs of claim investigation and handling, the costs of disputes over whether the insurance policies cover the claims, and the costs of defending policyholders in disputes with other PRPs or the

Table 4. Breakdown on Insurer Costs by Type of Expenditure

Type of cost	Share of total costs (percent)
Transaction costs	88
Claim investigation	9
Coverage disputes	42
Policyholder defense	37
Indemnity	12
Total	100

Source: Acton and Dixon 1992, 24.

government. We found that 88% of insurer outlays through 1989 were transactional in nature and that expenditures both on coverage disputes and policyholder defense were substantial.

Private-Sector Transaction Costs through 1991

Overall private-sector expenditures at Superfund sites are the sum of PRP and insurer expenditures.[6] In a 1994 study, we estimated combined private-sector outlays and transaction-cost share using results from the 1992 and 1993 studies, as well as data from EPA on overall PRP expenditures through 1991 (Dixon 1994). This estimation was based on three assumptions:

- Good estimates of the transaction-cost share of private-sector PRP and insurer expenditures through 1991 were 32% and 88%, respectively.
- During 1990 and 1991, insurer expenditures grew at the average rate observed in 1988 and 1989 for the four insurers (35%).
- Actual PRP outlays at NPL sites through 1991 equaled the value of PRP response settlements reported by EPA, once adjusted for likely cost overruns, an estimated private-sector efficiency advantage, and EPA cost recovery through 1991 (Dixon 1994, 63–66).

This last assumption deserves particular scrutiny: there is considerable uncertainty regarding how closely the negotiated PRP settlements through 1991 reflect true private-sector cleanup expenditures through 1991. On the one hand, the EPA figures do not include settlements negotiated by the states or work done by PRPs without a formal agreement.[7] On the other, these settlements are only commitments to pay: the actual outlays may come later. The EPA data, however, are the only data available.

Table 5 presents projected private-sector expenditures and transaction-cost share at NPL sites through 1991. To avoid double counting,

Table 5. PRP, Insurer, and Total Private-Sector Expenditures and Transaction-Cost Shares through 1991

	PRPs (1)	Insurers		Total (4)	Total private sector (1)+(3)
		To PRPs (2)	Not to PRPs (3)		
Outlays ($ millions)					
Cleanup	7,219	121	0	121	7,219
Transaction cost	3,397	186	700	886	4,097
Total	10,616	307	700	1,007	11,316
Transaction-cost share (percent)	32	61	100	88	36

Source: Dixon 1994, Table A.3.

total private expenditures excludes column 2.[8] Estimated private-sector outlays through 1991 were approximately $11 billion, with insurers accounting for about 10% of the total. Transaction costs were estimated to be 36% of overall private-sector expenditures. For comparison, EPA expenditures on the Superfund program over the same period were approximately $9.1 billion (Acton 1989, 31; U.S. EPA 1993, I-5).

Private-Sector Transaction-Cost Share at Cleanup Completion

So far I have discussed private-sector expenditures and transaction-cost share through 1991. However, it may be more important when evaluating Superfund's liability approach to consider the expenditures and transaction-cost share once cleanup has been completed at all NPL sites. At the eighteen study sites examined in the 1993 RAND study, we divided the cleanup process into three phases and found that transaction-cost share was lower in later stages of the cleanup process. Transaction-cost share fell from 51% of expenditures from the time of site discovery to the start of the first Remedial Investigation/Feasibility Study (RI/FS), and 39% from the start of the RI/FS to the start of the first remedial action, to 20% after the start of the first remedial action (Dixon, Drezner, and Hammitt 1993, 46). This suggests that the transaction-cost share through 1991 would overestimate the share of transaction costs in total expenditures when cleanup is complete.

Because cleanup is complete at relatively few sites, there is a great deal of uncertainty in the transaction-cost share at cleanup completion. To give some feel for what that share may be, in the 1993 study we projected transaction-cost share at completion for the eighteen study sites under three scenarios. To a baseline scenario we added both a scenario where ultimate site cleanup costs are 50% higher than expected in 1991 and a scenario where no more transaction costs are incurred after 1991.[9] This last scenario represented a lower bound for the ultimate transaction-cost share at the study sites. The resulting transaction-cost shares ranged from 19% to 27%, well below the 32% estimated for PRP expenditures through 1991 (Dixon, Drezner, and Hammitt 1993, 51).

Considerable uncertainty exists in the estimate for each scenario, as well as over which scenario best approximates the future. For example, using the upper and lower bounds of the 90% confidence intervals for the estimated transaction-cost share in each of the three cleanup phases causes the 27% share in the baseline scenario to range from 13% to 41%. What is more, the estimates for each scenario are based on the extrapolation of past results. Changes in the Superfund law or in program implementation could increase or decrease ultimate transaction-cost shares.

As more insurance claims are resolved, insurer transaction-cost share may also fall from the 88% observed through 1989. We found that the transaction-cost share for claims closed at the time of our 1992 analysis was 69%. We used this as an estimate of insurer transaction-cost share when all claims have been resolved, but it may be inaccurate for several reasons. First, the relatively small number of claims closed through 1989 may not accurately represent all insurance claims: they may, for instance, be the least controversial claims. Second, the ultimate insurer transaction cost will depend a great deal on how the insurance coverage cases are resolved. If courts rule in favor of PRPs on coverage cases, for example, large indemnity payments may dilute transaction costs, reducing insurer transaction-cost share.

Table 6 presents RAND's estimates of ultimate private-sector transaction costs when cleanup is complete at the eighteen study sites. Combining the 69% share for insurer expenditures with the 19% to 27% range for PRP expenditures results in an overall range of 23% to 31%.

Even though the transaction-cost share for expenditures at the eighteen study sites through 1991 may well be representative of the transaction-cost share for private-sector expenditures at all NPL sites through 1991, the projected transaction-cost share for the eighteen study sites at completion may not be representative of the share for all NPL sites at completion. Sites that had substantial expenditures through 1991 may not well

Table 6. Private-Sector Transaction-Cost Shares at the Eighteen Study Sites at Completion

Type of transaction cost	Transaction-cost share (percent of total costs)
PRPs	19–27
Insurers	69
Total	23–31

Source: Dixon 1994, Table 1.2.

represent the entire NPL, and thus the transaction-cost share at completion at all NPL sites may be different from those at the eighteen study sites.

In fact, some evidence suggests that the eighteen study sites are not representative of all NPL sites, even though they may be representative of sites that had substantial private-sector expenditures through 1991. Two key determinants of transaction-cost share are the number of PRPs at a site and the expected total cleanup costs. Table 7 presents the total expected cleanup costs stratified by the number of PRPs per site, first for the 428 NPL sites for which data were available in an EPA database and then for the 18 study sites.[10] These findings suggest that sites with many PRPs may be overrepresented in the eighteen study sites, implying that the ultimate transaction-cost share at all NPL sites may be lower than the range estimated for the eighteen study sites. We at RAND hope to use a more complete EPA database, which has become available recently, on NPL site characteristics to project our findings on transaction-cost share at completion to the entire NPL.

Government Transaction Costs

The transaction costs of local, state, and federal governments must be added to those of the private sector to determine the overall transaction costs generated by the Superfund process. Thirteen percent of EPA's 1993 $1.6 billion Superfund budget was for enforcement, clearly a transaction cost (U.S. EPA 1993, IV-2). However, it is not known what part of the remaining 87% of EPA's Superfund budget is transactional, and a comprehensive analysis of state and local spending has yet to be done.

TRANSACTION COSTS IN SUPERFUND VERSUS TORT LITIGATION

A liability approach is used in many other settings in the United States to resolve disputes and create appropriate incentives. In this section, I compare transaction costs in the Superfund process with those in another application of the liability approach—tort litigation.

The world of tort litigation is very different from Superfund litigation, both in terms of the law that is applied and the actions requested. Tort litigation draws mainly on common law while Superfund litigation is statutorily driven. Both sides in tort litigation are usually private parties, but in the Superfund context EPA is analogous to the plaintiff. Monetary damages are largely the goal of tort suits whereas PRP agree-

Table 7. Share of Total Expected Cleanup Costs by Number of PRPs at the Site

Number of PRPs at site	Sites in 1991 EPA survey (percent)	18 study sites (percent)
≤15	60	33
16–100	20	22
>100	19	44
Total	100	100

Source: Dixon, Drezner, and Hammit 1993, 29.

ments to do the cleanup are often EPA's goal. Nevertheless, even though tort limitation is very different from Superfund litigation, it does provide some reference point for evaluating Superfund transaction costs.

Superfund transaction costs are compared here with transaction costs for three different classes of torts: asbestos personal injury cases closed before August 1982; major U.S. aviation accident deaths between 1970 and 1984; and all tort litigation concluded in the United States in 1985. Studies conducted by the Institute for Civil Justice at RAND provide the data used for these comparisons.

Asbestos litigation closed through 1982 provides what many consider to be an upper bound on transaction costs relative to compensation for tort cases (Kakalik and others 1988, 86). This is due to several factors. First, at least through 1982, the defendants usually contested liability with the plaintiff. Second, defendants usually fiercely contest liability among themselves. Third, defendants contest the amount of the award. All this is similar to Superfund in that PRPs often fiercely contest not only the allocation of liability among themselves and their insurers, but also the type of remedy, which is analogous to contesting the size of the award in asbestos cases. In contrast to asbestos litigation, however, private-sector PRPs have largely given up disputing their liability with EPA.

RAND found that transaction costs accounted for 50% of total defendant expenditures on asbestos claims closed through 1982 (Kakalik and others 1988, 94).[11] Since this transaction-cost share is for closed claims, it is compared most directly to Superfund transaction-cost share at completion of cleanup. The share for asbestos claims is substantially higher than the 23% to 31% estimated share at completion for Superfund.

Airline accident litigation provides an example of a low transaction-cost share in the tort setting. In contrast to asbestos litigation, the defendants in airline litigation generally do not contest liability with the plaintiff and often simply agree among themselves over the allocation of liability. The major issue under dispute is the size of the award

(Kakalik and others 1988, 92). RAND found that transaction costs were 14% of total defendant expenditures for airline accident death cases between 1970 and 1984 (Kakalik and others 1988, 94). This is well below the estimated transaction-cost share at completion for Superfund cleanups.

Finally, a RAND study found that transaction costs accounted for 35% of defendant expenditures for all tort cases concluded in the United States in 1985 (Kakalik and others 1988, 92). This is slightly above the 23% to 31% range estimated for Superfund.

A more thorough analysis would compare Superfund transaction costs with those in other compensation schemes, such as the workers' compensation system, as well as with transaction costs generated by other approaches to national problems, such as traditional public works programs. A more thorough comparison would also need to compare benefits produced by the various schemes. For example, by requiring private parties to foot the bill, Superfund might keep cleanup costs below what they would be in a public works program. The transaction-cost share generated by the Superfund process appears to fall in the range of shares generated by the tort litigation examined here. However, this should not divert attention from the basic question: can the Superfund process be reformed to maintain the benefits of the program but at reduced transaction costs?

CONCLUSIONS

The data we at RAND have collected so far suggest that the approach adopted in the United States to clean up abandoned or inactive hazardous waste sites generates substantial transaction costs. We estimate the following.

- Transaction costs accounted for 32% of private-sector PRP expenditures through 1991.
- The overwhelming majority (88%) of insurer expenditures on insurance claims related to hazardous waste cleanups were transaction costs.
- The private sector incurred $11.3 billion in NPL-related expenditures through 1991, of which 36% were transactional in nature.

We have also found that transaction costs and transaction-cost shares vary considerably across firms and sites.
- Firm expenditures are higher but transaction-cost shares are lower for firms with larger volumetric shares.
- Transaction-cost shares are higher at sites with more PRPs.

- Transaction costs are higher but transaction-cost shares are lower at sites with greater expected cleanup costs.
- Transaction-cost shares are lower in later stages of the cleanup process.

A great deal of uncertainty remains on the magnitude of private-sector transaction costs, particularly on what the transaction-cost share will be when cleanup is complete. One of factors contributing to this uncertainty is the difficulty of collecting data from private-sector PRPs. PRPs are often unwilling to provide information on their cleanup and legal costs, even on a confidential basis. Even if more complete data were available on expenditures to date, a great deal of uncertainty would remain over the final transaction-cost share. This is because, first, relatively few sites have completed the cleanup process so far and, second, PRP strategy for cost recovery from other PRPs and insurers may change over time as cleanups near completion.

Superfund transaction-cost shares appear to fall in the range observed for common types of tort litigation. This should not divert our attention, however, from asking whether the program can be reformed or redesigned to maintain its benefits but reduce the transaction costs.

ENDNOTES

[1] The eighteen study sites were selected randomly from the set of NPL sites where we had reason to believe that there had been substantial private-sector expenditures through 1991.

[2] Of the 108 firms sampled at the eighteen study sites, 31% spent less than $1,000 each through 1991. At the other extreme, seven firms spent $14.7 million on average (Dixon, Drezner, and Hammitt 1993, 22).

[3] The *firm* characteristics were volumetric share (the proportion of the waste by volume at the site sent by the firm) and firm size. The *site* characteristics were the number of PRPs at the site, expected site cleanup costs, whether there was municipal involvement, and whether the government helped finance the remedy.

[4] Note that sites where there had been nontrivial PRP expenditures through 1991 need not look much like all sites on the NPL in 1991.

[5] Other site characteristics held constant in the analysis are: expected site cleanup cost, whether there is municipal involvement, whether PRPs alone or PRPs and the government together financed the cleanup, and cleanup stage.

[6] Private-sector expenditures also include expenditures by community groups that participate in the cleanup process and by reinsurers. Because these outlays were unlikely to be large relative to PRP and insurer expenditures through 1991, private-sector expenditures are excluded from the analysis.

[7]For example, a PRP may have cleaned up contamination on its own property on or near a site without formal agreement with EPA. During the course of our studies, we found several examples of this.

[8]Insurer outlays cannot simply be added to PRP outlays to determine total private-sector outlays because some insurer payments are made to PRPs (see Dixon 1994, 65).

[9]In the baseline scenario, we used the transaction-cost shares for each cleanup stage as estimated through 1991 and the expected total site cleanup costs as reported by EPA regional project managers in 1991.

[10]The EPA database on NPL sites alternatively is referred to as the CBO/GAO/RFF Survey or the SSRS Database. It contains data collected from EPA regional project managers on NPL site characteristics. The completeness and accuracy of some of these data are highly questionable. For example, information on both the number of PRPs and the expected cleanup costs was available for only 428 of the approximately 1,275 sites on the NPL. However, at the time of this writing, the SSRS was the most complete database publicly available. EPA provided the database to RAND in June 1993.

[11]The RAND study also reported plaintiff transaction costs, but these are not included here for the sake of comparison with Superfund. In the Superfund context, EPA is equivalent to the plaintiff in tort cases, and we do not know EPA's transaction costs.

REFERENCES

Acton, Jan A. 1989. *Understanding Superfund: A Progress Report.* RAND/R-3838-ICJ. Santa Monica: RAND.

Acton, Jan A., and Lloyd S. Dixon. 1992. *Superfund and Transaction Costs: The Experiences of Insurers and Very Large Industrial Firms.* RAND/R-4132-ICJ. Santa Monica: RAND.

Dixon, Lloyd S. 1994. *Fixing Superfund: The Effect of the Proposed Superfund Reform Act of 1994 on Transaction Costs.* RAND/MR-455-ICJ. Santa Monica: RAND.

Dixon, Lloyd S., Deborah S. Drezner, and James K. Hammitt. 1993. *Private-Sector Cleanup Expenditures and Transaction Costs at 18 Superfund Sites.* RAND/MR-204-EPA/RC. Santa Monica: RAND.

Hedeman, William N., Jonathan Z. Cannon, and David M. Friedland. 1991. Superfund Transaction Costs: A Critical Perspective on the Superfund Liability Scheme. *Environmental Law Reporter* 21 (7): 10,413–10,426.

Kakalik, James S., Elizabeth M. King, Michael Traynor, Patricia A. Ebener, and Larry Picus. 1988. *Costs and Compensation Paid in Aviation Accident Litigation.* RAND/R-3421-ICJ. Santa Monica: RAND.

U.S. EPA (Environmental Protection Agency). 1993. *Superfund Management Report.* December 6, 1993. Washington, D.C.: U.S. EPA.

8
De Minimis Settlements under Superfund: An Empirical Study

Lewis A. Kornhauser and Richard L. Revesz

This chapter reports the results of an empirical study of de minimis settlements entered on or before June 30, 1992. We analyzed the settlement documents and other site-specific information for virtually all of the de minimis settlements entered through June 30, 1992, as well as various databases prepared by the U.S. Environmental Protection Agency (EPA) that contain information about sites on the National Priorities List (NPL). We also interviewed the attorneys charged with primary responsibility for de minimis settlements at each of the EPA regional offices, as well as selected representatives of potentially responsible parties (PRPs) in Superfund actions, both de minimis and non–de minimis.

We focus on three principal questions: How many de minimis settlements were entered, as a percentage of the universe of sites for which such settlements would be appropriate? When, during the long process of cleanup, were the settlements entered? What variations were there in the terms of the settlements?

Unfortunately, our conclusions are far from encouraging. EPA has vastly underutilized de minimis settlements, using this tool in only about one-fifth of the sites likely to benefit from such settlements. Even in those instances, it has settled late in the cleanup process, after many years of legal wrangling have greatly reduced the benefits of settlement. It has also failed to follow its own policy of standardizing the form of the settlements, thus creating incentives for costly negotiations over the terms of de minimis settlements and for conflict between de minimis and non–de minimis defendants. Moreover, the variations in settlement practices cannot be explained on the basis that the de minimis program is administered in a decentralized fashion by the EPA regions (the agency has ten regional offices with geographic jurisdiction); there are substantial variations even within single regions.

The questions addressed in this chapter are important to an evaluation of the merits of the Superfund liability scheme, which has been criticized for the high transaction costs that it imposes on the affected parties. This criticism is most apt in the case of parties that bear a small share of the liability at a site. RAND recently studied transaction-cost shares (the ratio of a firm's transaction costs to the sum of its transaction costs and remediation costs) for various types of PRPs. It found that a firm's transaction-cost share increases significantly as its volumetric share at a site falls (Dixon, Drezner, and Hammitt 1993, 35–37). Thus, de minimis parties bear a disproportionate amount of transaction costs. This problem could be alleviated considerably by a well-functioning de minimis settlement program. Unfortunately, our study establishes that the program has been seriously underutilized and mismanaged.

NUMBER OF SETTLEMENTS

Through June 30, 1992, there were a total of seventy-six de minimis settlements involving forty-nine sites: sixty-seven were settlements with generators and/or transporters (that is, waste contributor settlements) and nine were settlements with current or prior owners of Superfund sites (that is, landowner settlements). There were forty-one sites with one or more waste contributor settlement, one site with both waste contributor and landowner settlements, and seven sites with one or more landowner settlement.[1]

Landowner settlements typically involve only one or two parties. In contrast, waste contributor settlements often involve dozens, sometimes even hundreds of parties. Thus, from the perspective of saving transaction costs, the latter are more significant. We therefore focus on the forty-two sites at which there was at least one waste contributor settlement and evaluate whether EPA has made sufficient use of de minimis settlements.

We believe that any site at which EPA has issued a Record of Decision (ROD) and that has at least twenty PRPs is a good candidate for a de minimis waste contributor settlement. While EPA needs to have sufficient information about cleanup costs to enter into a de minimis settlement, there is no plausible argument that the cost estimate in the ROD for the chosen remedial plan is inadequate for these purposes. Moreover, because of the typical distribution of percentage contributions among generators, a site with twenty parties is likely to have several that contributed less than 1% of the waste—the most commonly used cutoff for de minimis settlements. (We recognize that this measure is somewhat crude and hope that we will be able to refine it in subsequent work.)

The analysis of the EPA's Superfund databases reveals that 233 sites, out of approximately 1,200 sites on the NPL, meet these criteria. (If one assumes that only sites with RODs and at least fifty PRPs are good vehicles for waste contributor settlements, the number of eligible sites declines only to 149.) Thus, EPA has entered settlements in only about 18% of these sites. Moreover, every region has underutilized the tool of de minimis settlements, although there are important variations across regions. Table 1 shows, for each region, the number of sites that would qualify for de minimis settlements with waste contributors, the number of settlements actually entered, and the latter number as a percentage of the former.

The criteria that we used (issuance of the ROD and twenty or more parties) are likely to understate the number of sites that could benefit from de minimis settlements, suggesting that this settlement tool has been even more underutilized than Table 1 reveals. First, the completion of a ROD is not a prerequisite for de minimis settlements. While EPA needs to have sufficient information about the cleanup costs, such information can be estimated at earlier stages in the cleanup process. In fact, about 10% of the existing settlements were reached before the entry of a ROD. Second, even sites with fewer than twenty PRPs may be appropriate for de minimis settlements. About 30% of the sites with de minimis settlements did not have twenty or more PRPs, at least as recorded in the databases that we analyzed.

Table 1. Waste Contributor Settlements by EPA Region

EPA region	Qualifying sites (A)	Sites with settlements (B)	$[(B)/(A)] \times 100$ (%)
1	28	8	28.6
2	30	3	10.0
3	33	1	3.0
4	24	6	25.0
5	67	12	17.9
6	14	6	42.9
7	5	1	20.0
8	9	0	0.0
9	13	1	7.7
10	10	4	40.0
Total	233	42	18.0

TIMING OF THE SETTLEMENTS

The central congressional objective in designing the provisions for de minimis settlements was that EPA resolve the legal obligations of parties with a small share of liability as early as possible in the cleanup process. This different congressional treatment of the two groups stems from the different transaction-cost shares that they face. In this part of the chapter, we evaluate the extent to which EPA has settled with de minimis parties before it was able to do so with major parties, the timing of the settlements relative to the different stages of the cleanup process, and the relative use of the less cumbersome settlement instrument.

Pure versus Nonpure Settlements

To analyze the question whether EPA is, in fact, entering into de minimis settlements before it is able to resolve the liability of the major parties, we define five separate categories of de minimis settlements:

- Global—De minimis settlements that are part of a global settlement, pursuant to which the major parties undertake to perform a cleanup
- Prior global—Settlements with only de minimis parties, but which were preceded by a global settlement or other settlements with major parties
- Major parties—Settlements, other than global settlements, with both de minimis and major parties
- Prior major parties—Settlements with only de minimis parties, but which were preceded by nonglobal settlements with major parties
- Pure de minimis settlements

In this taxonomy, we define a global settlement to be one pursuant to which the major parties, or a group of major parties, undertake to perform the Remedial Design and Remedial Action (RD/RA) at the site.[2] In all but the fifth category, EPA is resolving the liability of the major parties either consecutively with, or earlier than, the liability of de minimis parties. Table 2 categorizes the forty-two initial de minimis waste contributor settlements at each site.

Table 2 shows that of these settlements, only fifteen (36%), were pure de minimis settlements.[3] In all other cases, the de minimis settlement did not occur before EPA was in a position to resolve the liability of the major parties. Most strikingly, in twenty cases (48%), the de minimis settlement occurred in conjunction with a global settlement.

We also looked at how this distribution of settlements varied by EPA region. This examination revealed that no region used pure de minimis settlements as its primary tool.

Table 2. Pure De Minimis Settlements and Settlements Including Major Parties

Type of settlement	Number of initial de minimis settlements
Global	20
Prior global	3
Major parties	3
Prior major parties	1
Pure de minimis	15
Total	42

Stage in the Cleanup Process

In this section, we categorize the de minimis settlements by reference to the stage in the cleanup process at which they were entered. We define three stages: Stage 1, pre-RI/FS (Remedial Investigation/Feasibility Study) settlements; Stage 2, settlements entered after the completion of the RI/FS but before the completion of the ROD; and Stage 3, settlements entered after the completion of the ROD.[4] In turn, for Stage 3 settlements, we indicate, by a number following the decimal point, the number of years, to the nearest whole year, that elapsed between the completion of the ROD and the entry of the settlement; thus, a settlement entered two years after the completion of the ROD would be indicated as being in Stage 3.2. Finally, the category labeled "Other" consists of removal, rather than remedial, actions.

Table 3 shows the distribution of the forty-two initial waste contributor settlements by reference to the stage in the cleanup process at which they occurred. We used as the date of the settlement the date on which EPA published notice in the *Federal Register*, rather than the time at which the settlement became final. (We lacked sufficient reliable data on the latter dates.) Thus, our analysis somewhat understates the delay in the entry of de minimis settlements.

The table shows that the vast majority of the settlements were concluded in the post-ROD period. Of the thirty-nine sites at which remedial action was contemplated, thirty-six settlements (92%) were entered after the signing of the ROD. Moreover, on average, considerable time elapsed between the signing of the ROD and the entry of the de minimis settlement. The average lag between the ROD and the first de minimis settlement was 1.8 years.

To put these numbers in some perspective, a RAND study found that after the listing of a site on the NPL, on average twenty months elapse until the beginning of the RI/FS, thirty-eight additional months until the issuance of the ROD and forty-three additional months until

Table 3. Timing of De Minimis Settlements

Stage of cleanup process	Number of contributor settlements[a]
1	1
2	2
3.0[b]	3
3.1	11
3.2	13
3.3	9
Other	3
Total	42

[a] These are the initial waste contributor settlements at each site, based on the date of the settlement (here defined as the date on which EPA published notice of the settlement).

[b] A number to the right of the decimal indicates years elapsed between the completion of the ROD and the entry of settlement.

the completion of the RD/RA—a total of 101 months from the NPL listing to the completion of the cleanup (Acton 1989, 16). Thus, considering only the first waste contributor de minimis settlement at a site, the post-ROD settlement is concluded eighty months after NPL listing.[5]

Next, we take into account the three post-RI/FS but pre-ROD settlements and the single pre-RI/FS settlement. We assume that these were entered at the midpoint of relevant interval. Then, the average de minimis settlement is concluded seventy-four months after NPL listing.

There are no studies about the pattern of expenditure of transaction costs throughout the cleanup process. If one were to assume that the expenditures are evenly distributed over time, by the time the average de minimis settlement is entered, the de minimis parties have expended approximately three-quarters of the total transaction costs that would be expended if the case did not settle until after the completion of the cleanup.[6]

Consider the effects of accelerating the settlement process. If, instead, the average de minimis settlement were halfway between the RI/FS and the ROD, only thirty-nine months would elapse between NPL listing and the settlement. The transaction costs expended, under the same assumption, would be only about half of what they are now. Even more dramatically, if the average de minimis settlement were entered halfway between the NPL listing and the RI/FS, only ten months would elapse between the NPL listing and the settlement. The result, under the assumption of uniform expenditure of transaction costs, would be transaction costs approximately eight times lower than they are now.

Next, we determine whether there are significant differences in the timing between pure de minimis settlements and other de minimis settlements, as defined in the previous subsection (global settlements, prior global settlements, major parties settlements, and prior settlements with major parties). Table 4 presents the relevant figures.

Table 4. Timing of Pure and Other De Minimis Settlements

Stage	Number of pure settlements at each stage	Number of nonpure settlements at each stage
1	0	1
2	1	1
3.0[a]	2	1
3.1	2	9
3.2	6	7
3.3	3	6
Other	1	2
Total	15	27

[a] A number to the right of the decimal indicates years elapsed between the completion of the ROD and the entry of settlement.

We use this table to compare the timing of nonpure and pure de minimis settlements. Taking into account all the settlements, 92% of the nonpure settlements (twenty-three out of twenty-five relevant observations) and 93% of the pure settlements (thirteen out of fourteen relevant observations) were entered after the signing of the ROD. Moreover, for post-ROD settlements, the average time elapsed between the settlement and the issuance of the ROD was 1.8 years in the case of both pure settlements and other settlements.

As these statistics show, the average pure settlement is not entered significantly earlier than the average settlement in one of the categories involving major parties. This conclusion is somewhat counterintuitive. Indeed, a logical hypothesis, which our analysis undermines, is that pure de minimis settlements would be entered early in the cleanup process, at a time when the information necessary to resolve the liability of the major parties is lacking.

We also examined how the timing of settlements varied across the regions with the highest number of settlements: Regions 1, 4, 5, and 6. For post-ROD settlements, the average time elapsed between the ROD and the entry of the settlement did not differ by more than half a year. The regional differences in timing are therefore not dramatic.

Settlement Instruments

We are interested in the choice between the use of consent decrees and administrative orders on consent because the latter do not require the assent of a third party (the court) and do not provide a formal forum for objections by nonsettlers. Thus, a settlement embodied in an administrative order is likely to be less cumbersome and, controlling for the timing of the settlement, is likely to involve a smaller expenditure of transaction costs on the part of the settling parties. Table 5 shows the use of the two types of instruments for pure and nonpure settlements.

Table 5. Distribution of Settlement Instruments by
Reference to the Role of the Major Parties

Instrument/type	Number of pure settlements	Number of nonpure settlements
Consent decree	6	23
Administrative order	9	4

The difference in the type of settlement instrument used is dramatic. Whereas in the case of pure de minimis settlements, only 40% (six out of fifteen) of the settlements used consent decrees, 85% (twenty-three out of twenty-seven) of the nonpure settlements used consent decrees. This difference is driven in part by the fact that, under the statute, global settlements must take the form of consent decrees.

We also studied the difference in the timing between settlements embodied in consent decrees and administrative orders for the forty-two initial waste contributor settlements at each site. For post-ROD settlements, the average lag between the ROD and the settlement was about 2.2 years for consent decrees and 1.3 years for administrative orders. Using the same assumptions as above, the average consent decree was entered eighty-five months after listing on the NPL, whereas the average administrative order was entered sixty-seven months after listing on the NPL. Thus, on average, consent decrees were entered one-and-a-half years later than administrative orders.

While we have not attempted to systematically study this causal connection, we believe that two factors may be at play. First, at a later stage, there is likely to have been more judicial involvement in the case, and a consent decree might therefore seem more appropriate. Second, the lesser degree of formality might speed up the settlement process. In summary, an independent cost of delaying the entry of de minimis settlements is the greater likelihood—indeed, the certainty in the case of global settlements—that the settlement instrument will be a consent decree rather than an administrative order.

TERMS OF DE MINIMIS SETTLEMENTS
FOR WASTE CONTRIBUTORS

We focus in this part of the chapter on three issues. First, we study the maximum volumetric contribution consistent with de minimis status. Second, we analyze the instances in which settlements can be reopened as a result of additional information about the parties' volumetric contributions. Finally, we study the premiums used in exchange for the waiver of reopeners for cost overruns and further response action (a more extensive cleanup than that contemplated at the time of settlement).

Volumetric Cutoff

Parties are generally eligible for de minimis settlements if their volumetric contribution is smaller than a given percentage of the total volume of wastes at the site. In analyzing the maximum volumetric contribution consistent with de minimis status for the seventy-six waste contributor settlements, we eliminated four categories. First, some settlements did not indicate the volumetric cutoff and merely restated the statutory standard that the contributions of the parties offered a settlement were minimal in comparison to the other hazardous substances at the facility. Second, some settlements expressed the maximum permissible volume in gallons but did not indicate the total number of gallons at the site; thus, we were not able to calculate the percentage contributions. Third, for sites with multiple settlements, we considered only one settlement, except in the one case in which the cutoffs were different. (The first waste contributor settlement in this case used a cutoff of 2.5% but the second used a cutoff of 0.85%.) Fourth, in one site, the two settling parties contributed a very small amount of waste—lower than any of the cutoffs in other cases, and the settlement merely said that these amounts satisfied the requirements for de minimis settlements.[7]

We were left with thirty-two observations, revealing a range of 0.1% to 10% These cutoffs therefore differ by a factor of 100. It is possible, however, that the 10% cutoff is somewhat aberrational because it involved a case in which the only settling party was the U.S. Air Force. Eliminating this settlement reduces the range to 0.1% to 2.5%—still a factor of 25.

We believe that two factors are the likeliest candidates to explain the differences: the EPA region that entered the settlement and the number of parties offered the settlement. As to the former, since the guidance documents are not clear as to what the cutoff should be, it is possible that different regions would adopt disparate policies.

As to the latter, the overall number of PRPs at an NPL site varies greatly, from almost one thousand to only one. A condition for de minimis settlements is that a sufficiently large proportion of the liability remain unresolved. For a given proportion of unresolved liability, the cutoff for de minimis status is likely to decrease as the number of parties increases. Thus, in general, one would expect that sites in which a larger number were offered a settlement would have a smaller cutoff.[8]

Table 6 shows for each different cutoff used, the number of settlements using that cutoff, the region in which the settlements were entered, and the range of the number of parties offered the settlement. For this purpose, we define the following ranges:
- Range 1: 9 or fewer parties
- Range 2: 10–19 parties

- Range 3: 20–49 parties
- Range 4: 50–99 parties
- Range 5: 100 or more parties

Where a region entered into more than one settlement with a given cutoff, the number of such settlements is indicated in parentheses.

It is interesting that fourteen of the thirty-two settlements (44%) used a single cutoff: 1%. One other cutoff (0.2%) was used by two settlements. All the other cutoffs were used by only one settlement. Six settlements used cutoffs above 1% and twelve used cutoffs below 1%.

To enable a clearer analysis of potential regional disparities in the choice of percentage cutoffs, Table 7 shows these cutoffs sorted by region. It also provides, for each settlement, the range of the number of parties.

Table 7 reveals that Region 4 used the same cutoff (1%) in four of its five settlements. These settlements covered the whole spectrum of the range of the number of parties. Region 1 used a single cutoff (also 1%)

Table 6. Percentage of Cutoffs in Ascending Order

Cutoffs (%)	Number of settlements	EPA region	Range of number of parties offered the settlement[a]
0.1000	1	5	5
0.1455	1	5	5
0.2000	2	1	5
		5	3
0.2500	1	6	5
0.3200	1	7	1
0.4499	1	5	5
0.5000	1	4	5
0.6000	1	5	1
0.7000	1	1	1
0.7140	1	1	5
0.8500	1	10	1
1.0000	14	1(4)	2, 5(3)
		2(2)	5(2)
		4(4)	1, 2, 4, 5
		5(3)	4, 5(2)
		6	[b]
1.2000	1	5	4
1.3600	1	10	4
1.6000	1	6	1
2.0000	1	10	4
2.5000	1	10	1
10.0000	1	10	1

Note: Where a region entered into more than one settlement with a given cutoff, the number of such settlements is indicated in parentheses.

[a]The range numbers used in this table are: Range 1: 9 or fewer parties; Range 2: 10–19 parties; Range 3: 20–49 parties; Range 4: 50–99 parties; Range 5: 100 or more parties.
[b]Missing information.

in four of the six settlements, and Region 5 used the 1% cutoff in three of its ten settlements. With the exception of these three cases, no region used a single cutoff in more than one case. With the exception of Region 4, it is difficult to discern from this table any consistent intraregional approaches to the determination of cutoffs.

Table 8 sorts the percentage cutoffs by the range of number of parties, to help assess the extent to which this factor accounts for the variability in cutoffs.

For each of the ranges with several settlements, there is a large variation in the cutoffs used. Perhaps most surprising is the variation in Range 5 (more than 100 parties offered the settlement), where cutoffs from 0.1% to 1% were used. The differences are not explained by different regional policies: in the latter range, Region 5 spanned the whole spectrum.

Table 7. Percentage of Cutoffs, Sorted by Region

EPA region	Number of settlements	Cutoffs (%)	Range of number of parties offered the settlement[a]
1	6	0.2	5
		0.7	1
		1(4)	2, 5(3)
2	2	1(2)	2, 5
3	0	N.A.	N.A.
4	5	0.5	5
		1(4)	1, 2, 4, 5
5	10	0.1	5
		0.1455	5
		0.2	3
		0.4499	5
		0.6	1
		0.714	5
		1(3)	4, 5(2)
		1.2	4
6	3	0.25	5
		1	[b]
		1.6	1
7	1	0.32	1
8	0	N.A.	N.A.
9	0	N.A	N.A.
10	5	0.85	1
		1.36	4
		2	4
		2.5	1
		10	1

Note: Where a region entered into more than one settlement with a given cutoff, the number of such settlements is indicated in parentheses. (N.A. = not applicable)

[a]The range numbers used in this table are: Range 1: 9 or fewer parties; Range 2: 10–19 parties; Range 3: 20–49 parties; Range 4: 50–99 parties; Range 5: 100 or more parties.
[b]Missing information.

Table 8. Percentage of Cutoffs, Sorted by the Range of the Number of Parties

Range of number of parties[a]	Number of settlements	Cutoffs (%)	EPA region
1	8	0.32	7
		0.6	5
		0.7	1
		0.85	10
		1	4
		1.6	6
		2.5	10
		10	10
2	3	1(3)	1, 2, 4
3	1	0.2	5
4	5	1(2)	4, 5
		1.2	5
		1.36	10
		2	10
5	14	0.1	5
		0.1455	5
		0.2	1
		0.25	6
		0.4499	5
		0.5	4
		0.714	5
		1(7)	1(3), 2, 4, 5(2)
[b]	1	1	6

Note: Where a region entered into more than one settlement with a given cutoff, the number of such settlements is indicated in parentheses.

[a]The range numbers used in this table are: Range 1: 9 or fewer parties; Range 2: 10–19 parties; Range 3: 20–49 parties; Range 4: 50–99 parties; Range 5: 100 or more parties.
[b]Missing infromation.

Reopener for Additional Information on Volume

The EPA guidance documents, which instruct the regional offices on the administration of the de minimis settlement program, set forth the conditions for a *reopener* of the settlement in the event that EPA obtains additional information about the settling party's volumetric contribution or the toxicity of its wastes. Settlements should generally include an "additional information" reopener, "which would allow the Government to seek further relief from any settling party if information not known to the Government at the time of the settlement is discovered which indicates that the volume or toxicity criteria for the site's de minimis parties are no longer satisfied with respect to that party" (U.S. EPA 1987, 15; 1989, 13–14). Such a reopener is not necessary if EPA believes that the probability of discovering new waste information about the site is negligible (U.S. EPA 1987, 15).

We focus in this section of the chapter only on the reopener for additional information on volume (rather than on toxicity). Surprisingly, under the approach of the guidance documents, this reopener is triggered by an actual volumetric contribution higher than the maximum volumetric contribution consistent with de minimis status, rather than a volumetric contribution higher than that reflected in the settling party's payment. Suppose that at a particular site, EPA defines parties that contributed less than 1% of the waste as de minimis. A party that, at the time of the settlement appeared to have contributed only 0.4% but that was later shown to have in fact contributed 0.8% would appear not to be subject to the additional information reopener, even though it should have paid twice as much as it did.

Of the forty-two sites with at least one de minimis settlement by waste contributors, an additional information reopener was used in thirty-seven (88%). Two of these sites (both in Region 1) involved multiple settlements in which the reopener was used for some settlements but not others.[9] In addition, the reopener was not used in five other sites (in Regions 1, 2, 5, and two sites in Region 9).

Even though the guidance documents prescribe a single formulation, the nature of the additional information that will trigger this reopener varies greatly across settlements. The following formulations are used to categorize these triggers and are used in Tables 9 and 10:

- Trigger 1: The volume contributed by the settling party is greater than the cutoff used to determine de minimis status—the approach of the guidance documents. (In two settlements that use this formulation, the actual cutoff is not indicated.)
- Trigger 2: The settling party's actual volumetric contribution exceeds the amount attributed to it at the time of the settlement, and the reopener is triggered. (In one case using this formulation, there is an exception for changes that result from recomputations of the contributions of the major parties.)
- Trigger 3: The settling party's actual volumetric contribution exceeds the amount attributed to it at the time of the settlement, but the reopener is not triggered; instead, the settling party pays an additional proportional amount.
- Trigger 4: The settling party's actual volumetric contribution significantly exceeds the amount attributed to it at the time of the settlement.
- Trigger 5: The settling party made material misrepresentations concerning its volumetric contribution.
- Trigger 6: The settling party made any misrepresentations concerning its volumetric contribution.

Some cases used multiple triggers.

Table 9 shows the regional distribution of the various triggers. For each site with multiple settlements, if the settlements include a reopener for additional information concerning volume, the triggers are identical. Thus, the table displays the trigger information for one settlement per site. The table identifies the triggers by the numbers set forth in the preceding paragraph, and indicates instances of more than one trigger per settlement.

The table shows that the most common trigger is the cutoff for de minimis status (Trigger 1)—the only one contemplated in the guidance documents. It was used on its own in twenty-two out of thirty-seven instances (59%) and in conjunction with another trigger in an additional eight settlements (22%). Five regions (Region 1, 2, 6, 9, and 10) used it for all their settlements, though Region 1 occasionally coupled it with another trigger. Two other regions (Regions 3 and 7), in contrast, never used it. The remaining two regions with waste contributor settlements (Regions 4 and 5) used it in some sites but not in others. Of these, Region 5 used the greatest number of different formulations. Its eleven settlements used five different types of triggers.

Premiums

The guidelines contemplate that reopeners for cost overruns and further response action can be waived in return for the payment of a premium. However, the guidelines do not specify the mechanism for determining the premium. (We do not deal here with premiums charged to parties because of their failure to accept an earlier settlement because the settlements often do not explain how they determine each offer.)

Table 9. Triggers for the "Additional Information" Reopener with Respect to Volume

EPA region	Trigger							
	1	2	4	1 and 3	1 and 5	1 and 4	1 and 6	1, 3, and 6
1	3	0	0	0	2	0	1	1
2	2	0	0	0	0	0	0	0
3	0	0	1	0	0	0	0	0
4	4	2	0	0	0	0	0	0
5	4	1	2	3	0	1	0	0
6	6	0	0	0	0	0	0	0
7	0	1	0	0	0	0	0	0
8	0	0	0	0	0	0	0	0
9	1	0	0	0	0	0	0	0
10	2	0	0	0	0	0	0	0
Total	22	4	3	3	2	1	1	1

Note: See text for a detailed description of the triggers.

DE MINIMIS SETTLEMENTS UNDER SUPERFUND   201

Unfortunately, our analysis of this issue is somewhat hampered party's payment and it is therefore not possible to ascertain whether the parties are paying a premium in addition to their volumetric share of the cleanup costs. A large majority of the forty-two settlements in our sample for which such information was available, covering twenty-five sites,[10] charged de minimis parties a premium. This approach was followed in thirty-eight settlements (90%). No premium was charged in four settlements, involving three sites. In two of these, the parties either had committed themselves to pay a percentage of the future costs or faced reopeners for cost overruns and further response action. In the other two cases (involving a single site in Region 1) the parties had limited solvency and the settlement took account of this fact.

Of the thirty-eight settlements in our sample for which we know that EPA charged a premium, information about how the premium was computed cannot be discerned in eight settlements, covering six sites. As to the other settlements, there is an important inconsistency in the approach followed. Some charge as a premium a percentage of the total cleanup costs at the site whereas others charge a percentage of only the future cleanup costs at the site.[11]

In nine of the remaining thirty settlements, involving six sites, the premium was charged only on the estimated future costs. In seven settlements, a fixed percentage was charged. These premiums ranged from 75% of the estimated future costs to 250%.[12] In all of these cases, EPA waived the reopeners for cost overruns and further response action. In the other two settlements, the settling parties were given the choice between a premium of 50% with reopeners for cost overruns and future response action or a premium of 150% without these reopeners.

In one of the thirty settlements, the premium was a percentage of past costs (10%) and a different percentage of future costs (50%). In this settlement, EPA waived the reopener for cost overruns, but retained the reopener for further response action.

In twenty settlements (ten sites), the premium was charged on the basis of total costs: past costs plus expected future costs. In eighteen of these cases (eight sites), a fixed percentage was used: these ranged from 20.95% to 210%.[13] Except in two of these settlements (one site), EPA waived the reopeners for cost overruns and further response action. In the remaining case, it waived only the cost overrun reopener. The premiums charged in these settlements were 20.95% and 25%. Eliminating them changes the lower end of the range to 53.33%.

One of the twenty settlements charged a premium of 100% to parties that had contributed less than 1% of the wastes and 50% to parties that had contributed more than 1%. Another settlement charged a premium of 60% and 200%, respectively, for costs involving the first and second operable units (the elements—such as groundwater treatment or soil removal—into which cleanups are sometimes divided).

We now seek to determine whether this large variation is attributable to different policies on the parts of the EPA regions or whether it is attributable to different stages in the cleanup process. With respect to the latter factor, one would expect that uncertainty with respect to cleanup costs would decrease at later stages in the cleanup process and that smaller premiums would then be required. We define the following ranges of premiums, which also are used in Table 10:

- Range 1: less than 100%
- Range 2: greater than or equal to 100% and less than 200%
- Range 3: greater than or equal to 200%

For each range we use the letters F and T to denote whether the premium is charged on future costs or total costs, respectively.

For consistency, we include only settlements that waived the reopeners for both cost overruns and further response action. In the case of the settlements with variable percentages, we use the higher in the case of different percentages based on the contributions of the parties (because the smaller parties, which paid a higher percentage, were far more numerous) and the lower in the case of different percentages for operable units (because we use the date of the first ROD to determine the site stage in the cleanup process, and the uncertainties are less as time passes after the signing of the ROD). Where the number of sites is different from the number of settlements, we indicate it following a slash.

Table 10 shows the regional distribution of premiums. It shows that, for settlements that charged a premium on total costs, the most prevalent range was Range 1. It was used in eleven out of eighteen settlements (61%), involving four out of ten sites (40%). In five out of the eighteen settlements, (28%), covering four out of ten sites (40%), Range 2 was used. Only two settlements (11%), involving two sites (20%), used Range 3.

In contrast, for settlements that charged a premium only on future costs, the most prevalent range was Range 2. It was used for seven out of nine settlements (78%), covering five out of seven sites (71%). The lower and higher ranges were used only in one settlement each.

With respect to intraregional inconsistencies, we can make two observations. First, one region (Region 5) charged a premium on the future cost component in some settlements and on total costs in other settlements. Second, the two regions with the largest numbers of settlements (Regions 1 and 5) exhibited considerable variation in the amount of premiums charged.

Table 11 shows the relationship between the premium charged and a site's stage in the cleanup process. The table does not suggest that higher premiums are charged earlier in the cleanup process, when the uncertainties might be greater. For example, the 1T, 2T, and 3T premium

Table 10. Regional Distribution of Premiums

EPA region	Premium range[a]					
	1F	1T	2F	2T	3F	3T
1	0	6/2	0	2	0	0
2	0	0	0	0	0	0
3	0	4/1	0	0	0	0
4	1	0	1	0	1	0
5	0	0	5/3	2/1	0	2
6	0	1	0	0	0	0
7	0	0	1	0	0	0
8	0	0	0	0	0	0
9	0	0	0	0	0	0
10	0	0	0	1	0	0
Total	1	11/4	7/5	5/4	1	2

Note: A slash appears in cases where there are multiple settlements at a single site. The number before the slash indicates the number of settlements and the number after the slash indicates the number of sites.

[a]Range 1, less than 100%; Range 2, greater than or equal to 100% and less than 200%; Range 3, greater than or equal to 200%. For each range, the letters F and T denote whether the premium is charged on future costs or total costs, respectively.

ranges each contain sites in Stages 3.0 and 3.3. Moreover, for premiums levied only on the future cost component, the two sites in Stage 3.3 are in Range 1 and Range 3, respectively, whereas all of the sites in Range 2 are in earlier stages of the cleanup process.

SURVEYS OF EPA REGIONAL OFFICES AND PRIVATE ATTORNEYS

During May and June 1992, we administered a detailed questionnaire to attorneys in the ten EPA regional offices. In June and July 1992, we conducted a survey of attorneys who had represented de minimis and non–de minimis PRPs at sites in which de minimis settlements were reached. We focused on the methods used to initiate and encourage de minimis settlements, the settlement approaches used by EPA, the role of the major parties with respect to de minimis settlements, the timing of the de minimis settlements, the criteria used to determine de minimis status, and the nature of reopeners and premiums. This part of the chapter sets forth the major findings of both surveys.

Methods to Initiate and Encourage De Minimis Settlements

We asked about the methods that the regions currently use to initiate de minimis settlements and to encourage the formation of de minimis groups that will present settlement offers to EPA. The unanimous

Table 11. Relationship between Premiums and Stage in the Cleanup Process

Stage	Premium range[a]					
	1F	1T	2F	2T	3F	3T
2	0	0	1	0	0	0
3.0[b]	0	2/1	0	1	0	1
3.1	0	2/1	2	2	0	0
3.2	0	1	2/1	1	0	0
3.3	1	6/2	0	1	1	1
Other	0	0	2/1	0	0	0
Total	1	11/5	7/5	5	1	2

Note: The number of sites is different from Table 10 because, for one site, the settlements took place at different stages in the cleanup process and are entered separately in Table 11. A slash appears in cases where there are multiple settlements at a single site. The number before the slash indicates the number of settlements and the number after the slash indicates the number of sites.

[a]Range 1, less than 100%; Range 2, greater than or equal to 100% and less than 200%; Range 3, greater than or equal to 200%. For each range, the letters F and T denote whether the premium is charged on future costs or total costs, respectively.

[b]The number to the right of the decimal indicates years elapsed between the completion of the ROD and the entry of the settlement.

response was that EPA generally does not take the lead and prefers to wait for the PRPs to organize themselves into groups and to approach EPA if they are interested in exploring the possibility of a de minimis settlement. None of the regions had a standard policy for initiating or encouraging de minimis settlements.

Some regions, however, do take some initiative. Not surprisingly, Regions 1 and 5, which have entered into the largest number of de minimis settlements, also appear to make the most efforts in this regard.

Region 1 stated that it generally has a "kick-off" meeting with PRPs after the RI/FS is completed and a week or two before sending the special notice for RD/RA, which triggers a period for negotiation. At the kick-off meeting, Region 1 will indicate that it envisions a de minimis proposal at the site and that it will accept de minimis proposals from the PRPs. In addition, Region 1 usually presents de minimis parties with a settlement offer at the same time that it enters into RD/RA negotiations with the major parties. Its preference, nonetheless, is to let the major PRPs and the de minimis PRPs work out a settlement among themselves, thus saving EPA's resources.

In one recent case, Region 3 mailed EPA's guidance concerning de minimis settlements to all PRPs and requested that PRPs seeking de minimis status explain why they qualified. Region 3 then made a determination as to which parties met the requirements for de minimis treatment and issued a unilateral order to the major PRPs at the site to perform the necessary remedy. Negotiations with the de minimis PRPs followed.

Region 5 sometimes initiates de minimis discussions after it obtains the waste-in information, which is a list of PRPs and their respective waste contributions. In such cases, it informs the relevant PRPs that they are candidates for de minimis settlements and often makes formal settlement proposals. For the majority of de minimis settlements that are not part of global settlements, Region 5 has started the discussions: it will circulate a draft de minimis proposal in cases in which there is no organized group of de minimis PRPs, but in which there is a large number of such parties.

Region 9 stated that, in cases for which it believes a de minimis settlement is appropriate, it will suggest in its general notice letters that PRPs consider organizing a de minimis group. In one case involving thousands of PRPs, it is planning to include a return postcard in the first general notice that it will send all PRPs. The card has two boxes, one asking whether the PRPs are interested in organizing a de minimis committee and the other asking whether, if the PRP is not interested in organizing a committee, the PRP would be interested in participating in a de minimis committee. The postcard requests a response within thirty days.

Region 10 indicated that it had recently begun to ascertain when de minimis settlements might be appropriate. In such cases, it notifies the PRPs and asks them to provide information in support of their claim for de minimis status and to indicate whether they are interested in a de minimis settlement.

The view of the attorneys for PRPs was that de minimis discussions are generally initiated by the PRPs and not by EPA. Several attorneys indicated that, despite pressure from EPA headquarters, the regions are not particularly interested in pursuing de minimis settlements. Several also urged EPA to become more involved in the de minimis settlement process; one of them advocated the use of a system of incentives and disincentives to induce the regions to comply with the guidelines of EPA headquarters.

One attorney complained that the settlement process would be greatly aided if EPA circulated the waste-in information earlier in the process. She stated that the de minimis parties are often ready to cash-out early, but can proceed only at EPA's slow pace. She also noted that because, in general, there is little reliable information about each PRP's volumetric contribution, it would be prohibitively expensive for the de minimis parties to prepare the waste-in information themselves.

Another attorney noted that the process of formation of de minimis committees was itself haphazard, as it requires a party with a great deal of commitment and a willingness to expend significant resources. He added that, among small parties, there is often no party that wishes to take a leadership role, and he suggested that EPA fill the vacuum.

Settlement Approaches

We asked whether the regions employ a standard settlement form different from the EPA model form contemplated by the guidance documents and whether they employ any distinctive approaches. There were several important findings in response to this set of questions.

First, despite the clear congressional interest that de minimis settlements be entered before EPA is in a position to resolve the liability of the major parties, the regions expressed a strong and almost unanimous preference for global settlements because they conserve administrative resources and make it more likely that the major PRPs will undertake the cleanup. The attorneys for PRPs confirmed that EPA preferred this approach.

On this question, Region 1 indicated that if its negotiations with the major parties do not succeed, it will issue a de minimis administrative order on consent on a take-it-or-leave-it basis. It will not engage in negotiations.

Region 5 encourages the de minimis parties to deal directly with the major parties. For example, the Superfund statute sets forth special notice procedures by which the administrator can trigger a period of negotiation with PRPs; during this period, and thereafter if the PRPs present a good-faith proposal for performing or financing a cleanup, the administrator may not commence a cleanup action or require the PRPs to do so. Region 5 takes the position that, following a special notice letter, a separate offer for a de minimis settlement is not a good faith offer; at this stage, the de minimis proposal should come as part of an agreement (by the major parties) to perform the remedy. It will deal directly with the de minimis PRPs only if the major parties are being uncooperative.

Second, while the regions generally indicated that they use the model agreement, they make adjustments on a site-specific basis and engage in negotiations over at least some of the terms. Not surprisingly, several regions indicated that de minimis settlements place a high burden on EPA's managerial resources. Only Regions 1 and 5 (those with the highest number of settlements) stated that at present they proceed on a take-it-or-leave-it basis.

In particular, Region 5 stated that when it circulates a draft of a settlement, it asks for written comments (by the de minimis and major parties) and indicates that it may respond unilaterally to any of the comments that it receives. It also states, however, that the cost terms are not negotiable. After reviewing the comments, it sends out a final settlement offer on a take-it-or-leave-it basis.

Role of the Major Parties

We inquired about whether non–de minimis parties typically object to the entry of de minimis settlements, the types of objections that they raise, and whether such complaints pose a significant threat to the de minimis settlement program. There was virtual consensus that the major parties often create significant roadblocks to de minimis settlements. The areas of conflict include the criteria used to define de minimis parties, the reliability of the waste-in lists, the amount of the premium, and, in particular, the uses of the proceeds of the de minimis settlement.

It is clear from the responses that the major parties see themselves involved in a zero-sum game with the de minimis parties: the greater the de minimis settlement, the less that the major parties will have to pay. Moreover, Region 1 noted that, except in the case of global settlements, major parties do not favor de minimis settlements because they believe that they can do better by pursuing actions for contribution. Their opposition can lead to a substantial delay of the de minimis settlement and a substantial drain of EPA resources. While it does not appear that they have been successful at blocking de minimis settlements once negotiations were underway, they seem to have diminished the appetite of several of the regions for pursuing such settlements.

The most important specific issue to be raised by the responses is that the major parties have a strong preference for global settlements that include a de minimis component, rather than de minimis settlements that occur earlier. Region 5 indicated that in the former case the proceeds of the de minimis settlement are given to the major PRPs as seed money for the site remedy, whereas under the latter case, they are placed in the Superfund.

It would seem that the effect of turning the proceeds from de minimis parties to the major PRPs in a global settlement would be to leave unpaid a greater part of EPA's past costs. The major PRPs that undertake the cleanup might believe, however, that EPA will seek these costs from nonsettling parties, if there are any, or that it might be more willing to compromise them in exchange for a cleanup of the site. Background conversations with EPA officials revealed that such compromises do, in fact, take place.

The PRPs noted substantial tension between de minimis and major parties. One attorney stated that this tension is fueled in part by the lack of sufficiently specific guidance on de minimis settlements. He stated that, if such guidance existed, the major PRPs would not demand that the de minimis parties pay exorbitant premiums.

Several attorneys indicated that the major parties generally have the capacity to derail de minimis settlements. They added that, as a result, the regions are reluctant to become involved with de minimis settlements because of the risk that they will derail a settlement with the major parties.

Timing of the De Minimis Settlements

We asked a series of questions concerning the stage of the cleanup process during which the regions pursue de minimis settlement. There was disagreement about the desirability of pre-ROD settlements. There were also reports that EPA headquarters has begun to pressure the regions into concluding de minimis settlements at earlier stages of the cleanup process.

Region 2 indicated that the lack of concrete estimates for the remedial costs should not preclude the entry of a de minimis settlement, because the matter can be addressed through the use of an appropriately high premium.

Region 5 stated that in the past it had not done de minimis settlements until the RI/FS was almost complete, but that it is currently attempting to proceed earlier (even before the RI/FS is started) in response to pressure from EPA headquarters. In the case of global settlements, however, it cannot proceed until the ROD is complete, since global settlements are part of an RD/RA consent decree and the special notice for negotiations concerning the RD/RA is not issued until after the ROD stage.

Despite its willingness to consider early settlements, Region 5 said that complaints about waste-in lists are frequent and that by the time they are resolved the RI/FS is well along. At that point, Region 5 has a mind-set that since global settlement negotiations may take place only one year later, it might not make sense to pursue a separate de minimis settlement. It appears clear that Region 5 will not take the initiative in this regard if the ROD will soon be issued.

In contrast to the view that early de minimis settlements are possible, Region 1 stated that most of the information about waste-in contributions and cost estimates for the site remediation is available only at or around the time that RD/RA negotiations commence, and that earlier settlements are therefore undesirable. As a result, it favors global settlements.

Regions 3, 7, and 10 stated that until the ROD is completed, the list of PRPs is not sufficiently accurate and the estimates of cleanup costs are too unreliable to permit it to prepare a de minimis settlement.

Region 10 added, however, that recently it has begun to attempt earlier settlements in response to pressure from EPA headquarters for the entry of de minimis settlements before the completion of even the RI/FS. Region 9 also reported the interest of EPA headquarters in earlier settlements but noted that the regions do not know what the remedial costs will be. It indicated that for early settlements to be possible, reliable methods for estimating costs would have to be developed.

De Minimis Status

We asked the regions whether they apply any specific criteria to determine a PRP's eligibility for de minimis status. The responses revealed that these determinations are made on an ad hoc basis. Although the regions purport to be guided by the EPA documents, these, as we have indicated above, do not make more concrete the statutory requirement that the amount of hazardous substances contributed be "minimal."

Region 1 stated that it has used cutoffs of 0.5%, 1%, and 2%, and that in general it will not use cutoffs larger than 2%. Region 3 responded that in one case it used a 2% cutoff, but based it on the contributions of only the identified PRPs. Region 5 reported that it looks for a "clean break" in the waste-in list, to avoid the argument that its cutoff is arbitrary. It also indicated that de minimis parties tend to comprise 15% to 30% of the total volume at the site. Region 8 reported that in one case, one of the criteria for de minimis status was the accuracy of the information submitted in response to EPA's requests. Region 10 stated that a party that contributed around 1% will generally be considered de minimis. It added, however, that it might accord de minimis status to a party that contributed 4% if, for example, the remaining parties contributed around 20%.

One private attorney stated that in some cases EPA appears to have a predetermined cutoff in mind when it considers a de minimis settlement proposal. He added that EPA is reluctant to reveal when it is applying predetermined criteria or what terms might be negotiable.

Reopeners and Premiums

We inquired generally about the regions' use of reopeners in de minimis settlements. Every region appears to use the additional information reopener, but many favor releasing the cost overrun and further response action reopeners in exchange for a premium.

For example, Region 1 stated that the purpose of de minimis settlements is to extinguish the liability of de minimis parties. It views that cost overrun reopener as inconsistent with this objective. Region 5 stated that it tends to give PRPs a menu of choices—that is, they can pay a higher premium and not be subject to certain reopeners.

With respect to the premium charged in exchange for the cost overrun and further response action reopeners, we asked whether the regions had used any standardized guidelines. There appears to be no established procedure for setting these premiums; the regions indicated that the premiums are determined in a site-specific manner, although they could not articulate a protocol for how they do this.

CONCLUSION

Our empirical study offers clear answers to the three questions we posed at the outset. First, de minimis settlements have been underutilized as a settlement tool. EPA has invoked them at roughly 20% of the sites at which they might be used.

Second, even at those sites at which de minimis settlements have been employed, they have been entered into only late in the process, often in conjunction with settlements with major parties, and generally using the most formal settlement tools. This use is at odds with the congressional intent for the use of de minimis settlements.

Third, important terms of the settlements, including the triggers for some reopeners and the premiums charged for them, vary needlessly across sites. Some terms obviously should be site specific, such as the estimated cleanup costs; the premium, which is a function of the uncertainty surrounding these costs; and the cutoff for de minimis status, which is a function of the distribution of volumetric contributions. Even these terms, however, ought to be determined on the basis of a standard protocol. Other terms, such as the reopener for additional information, ought not to vary across sites. As we explain below, we believe that the unnecessary lack of uniformity is a cause of many of the difficulties exhibited by the de minimis settlement program.

Our interviews with attorneys at EPA's regional offices revealed a powerful explanation for the underutilization of de minimis settlements. The unanimous response was that EPA does not generally initiate discussions concerning such settlements. Similarly, EPA does not make an effort to encourage the formation of de minimis groups that could present settlement offers. Instead, EPA's position is largely reactive: it considers settlement proposals made by the de minimis parties. The problem with this approach is that it is often quite difficult for de minimis parties to organize a de minimis committee, so that in many cases, if EPA does not take the lead at offering de minimis settlements, such settlements will not occur.

Our interviews with attorneys at the regions also revealed that major parties tend to oppose de minimis settlements (except when they are part of global settlements) and that their opposition can lead to a substantial delay of the de minimis settlement and a drain of EPA's resources. While it does not appear that major parties have been successful at blocking de minimis settlements once negotiations were underway, those parties seem to have diminished the interest of several regions in pursuing such settlements.

We believe that this wrangling would be significantly reduced if de minimis settlements exhibited more standardized terms. Indeed, the current lack of uniformity gives the major parties a strong incentive to become involved in the de minimis settlement process. Because the liability of the major parties is reduced by the amount of the settlement, they benefit if they can persuade EPA to insist on a higher premium or more liberal reopeners. Such behavior would be less prevalent if the regions exercised less discretion on these matters.

Post-Study Developments: EPA Guidance Documents

Developments subsequent to the completion of our study—though responsive to some of the concerns expressed both here and in a prior report that we prepared for the Administrative Conference of the United States—do not materially change the situation. Two guidance documents, published in June 1992 and July 1993, respectively (U.S. EPA 1992, 1993b), state that the regions *may* assist in the formation of de minimis groups and should consider offering individual de minimis settlements without waiting for de minimis groups to form.

Unfortunately too much discretion continues to be vested in the regions, which, as we have seen, generally have not been eager to use this settlement tool. Instead of merely authorizing regions to seek de minimis settlements, EPA should establish procedures for negotiating de minimis settlements as a standard practice at all multiparty Superfund sites involving de minimis parties. As soon as EPA has sufficient volumetric information to make possible a de minimis settlement, it should advise any PRPs deemed to have made de minimis contributions about the availability of de minimis settlements and explain their general features, and as soon as there is sufficient information to estimate the cleanup costs at the site, EPA should circulate to these parties a draft settlement agreement.

With respect to the delay in the entry of settlements, the June 1992 guidance document urges the regions to consider pre-ROD de minimis settlements, before there is site-specific information about the cleanup costs. Instead of waiting for such information to become available, it asks the regions to identify similar sites at which cleanup is ongoing and to use the costs at those sites as a proxy. Alternatively, the June 1992 guidance document authorizes the regions to establish costs per unit of waste treated under different remedial technologies (U.S. EPA 1992). The July 1993 document indicates that a waste-in list is no longer a prerequisite for de minimis status; instead, a region can assess an individual PRP's waste contribution relative to the volume of waste at the site, without waiting for information about the contribution of all PRPs (U.S. EPA 1993b).

While these steps are positive, they are not sufficient. We believe that primary responsibility for determining the cleanup costs of different types of sites and the unit costs of various remedial technologies should be vested with EPA's headquarters rather than with the regions. It is not practical for a region to confine itself to its own sites in determining the costs of similar cleanups, as the inventory of comparable sites that have progressed sufficiently in the cleanup process may be small or nonexistent. Unless serious attention is given to this issue, it is unlikely that a significant proportion of de minimis settlements will be entered prior to the signing of the ROD.

Moreover, the current delay in the entry of settlement cannot be attributed to the delay in obtaining waste-in lists. As we have indicated, on average de minimis settlements are not entered until approximately two years following the issuance of a ROD, whereas waste-in lists tend to be available considerably earlier in the cleanup process.

ACKNOWLEDGMENTS

We initially undertook this project as consultants to the Administrative Conference of the United States (ACUS). An earlier version of this chapter was presented as our final report to ACUS and led that organization to adopt a series of recommendations to EPA for the improvement of the process of de minimis settlements under Superfund. Following the completion of our report to ACUS, we performed additional empirical work, which was funded by a grant from EPA's Office of Exploratory Research (contract #818460-01-1). The views presented in this chapter are those of the authors and not those of the members of ACUS or its committees, or of EPA.

We also acknowledge the generous financial support of the Filomen D'Agostino Greenberg and Max E. Greenberg Research Fund at the New York University School of Law and the help of the Information Network for Superfund Settlements, which was an important source of the settlement documents that we analyzed. We are grateful for the able research assistance of Michael Anastasio, Jeffrey Benz, Kent Chen,

Sarah Flanagan, and Jeffrey Spear. Vicki Been, Colleen Shannon, and Richard Stewart gave us valuable comments on an earlier draft.

ENDNOTES

[1] EPA recently released a document showing that, through October 1993, EPA had entered a total of 125 de minimis settlements at 78 sites. Of these, 112 were with waste contributors and 13 were with landowners (U.S. EPA 1993a).

[2] We do not include in this category settlements in which the major parties undertake to perform only the Remedial Investigation/Feasibility Study (RI/FS). We do include one settlement in which the major parties undertook only the operation and maintenance, and one in which EPA performed the remainder of the RD/RA.

[3] Note, moreover, that our methodology may overestimate the number of pure de minimis settlements. Conceivably, some of them may have been preceded by settlements with major parties, but this fact could not be ascertained from the settlement document.

EPA (U.S. EPA 1993a, 6) reports that only 31% of the settlements were global, whereas 69% were with de minimis parties only. The latter statistic is misleading, because it includes, as well as pure de minimis settlements: prior global settlements, prior settlements with major parties, and concurrent non-global settlements including major parties.

[4] Two of the sites in our sample had more than one physical location, and these locations were at different stages of the cleanup process. One of the sites had more than one operable unit, and these were at different stages of the cleanup process. In these two cases, we considered the latest stage that the site had reached.

[5] In the text, we use the average statistics about the time that it takes sites to pass through the various stages of the cleanup process as a way of establishing a rough comparison of the transaction costs entailed in pre-ROD and post-ROD settlements. We are currently redoing the analysis by considering the actual date on which each site was listed on the NPL.

[6] Of course, additional transaction costs would be expended if there were litigation rather than settlement following the conclusion of the cleanup. It is likely, however, that the discovery costs undertaken during the cleanup would be greater than the costs of the trial.

[7] The percentage contributions were 0.071% and 0.051%, respectively; it is possible in this case that EPA would have been also willing to offer de minimis settlements to parties with higher volumetric contributions.

We did include, however, a settlement that stated the amount contributed by the single settling party (0.7%) and indicated that this amount was consistent with de minimis status. The reason was that, in this case, the contribution was higher than cutoffs for other settlements.

[8] The problem is actually somewhat more complicated. It is possible that, for some settlements, EPA has extended the offer only to a subset of the qualifying par-

ties. In those cases, it would be more appropriate to consider the number of parties below the cutoff. The full analysis of this question would require the examination of the waste-in lists, which are almost never appended to the settlements.

It might also be appropriate to consider, instead, the total number of parties at a site. We attempted to do this, using a database prepared by EPA that contains, for each site, the number (and identity) of the parties that received formal notice of their potential liability. We noticed, however, that in several cases this number was smaller than the number of parties offered a de minimis settlement, indicating that EPA does not send notices to all PRPs. We are currently attempting to resolve this problem.

We also had difficulty, in some cases, in distinguishing between the number of parties that were offered the settlement and the number that accepted the settlement. In some cases, an appendix to the settlement lists all the parties below the cutoff and there is therefore no ambiguity. In other cases, however, the body of the settlement contains the identity of the settling parties. It is possible, in these cases, that other parties may have been offered the settlement as well and rejected it, but there is no way to tell from the settlement documents. In those cases, we used the number mentioned in the settlement. This problem is present primarily in cases involving a small number of parties.

[9] One of these instances involved two settlements with parties having limited solvency. Perhaps this factor accounts for the different treatment of these settlements. The explanation in the other instance is less easily discernible.

[10] At one site (in Region 1), information about the use of a premium was missing for two settlements, a premium was charged in four settlements, and no premium was charged in two settlements.

[11] It is possible, of course, that in the former case the premium is computed on the future cost component only, but that it is then expressed as a percentage of total costs. There is nothing in the settlements, however, to indicate that EPA is following such an approach. In fact, the settlements in which the premium is based on the total costs generally do not separate the future cost component.

[12] This large premium might be explained, at least in part, by the fact that the settlement applied only to a party with a minuscule contribution and that the party's total payment was only $510. The next highest premium in this category was 180%.

[13] It is possible, though we could not tell from the text of the settlement, that this premium includes a 100% premium for failing to have settled earlier. Indeed, this settlement is the third at a site, and the previous two were charged a premium of 110%. Even if this were the case, however, the range would not change dramatically; the next highest premium is 200%.

REFERENCES

Acton, Jan Paul. 1989. *Understanding Superfund: A Progress Report*. RAND/R-3838-ICJ. Santa Monica: RAND.

Dixon, Lloyd S., Deborah S. Drezner, and James K. Hammitt. 1993. *Private-Sector Expenditures and Transaction Costs at 18 Superfund Sites*. RAND/MR-204-EPA/RC. Santa Monica: RAND.

U.S. EPA (Environmental Protection Agency). 1987. *Interim Guidance on Settlements with De Minimis Waste Contributors Under Section 122(g) of SARA*. (OSWER Directive #9834.7). Washington, D.C.: U.S. EPA, Office of Solid Waste and Emergency Response.

―――. 1989. *Methodologies for Implementation of CERCLA Section 122(g)(1)(A) De Minimis Waste Contributor Settlements*. (OSWER Directive #9834.7-1B). Washington, D.C.: U.S. EPA, Office of Solid Waste and Emergency Response.

―――. 1992. *Methodology for Early De Minimis Waste Contributor Settlements Under CERCLA Section 122(g)(1)(A)*. (OSWER Directive #9834.7-1C). Washington, D.C.: U.S. EPA, Office of Solid Waste and Emergency Response.

―――. 1993a. *The First 125 De Minimis Settlements: Statistics from EPA's De Minimis Database*. Washington, D.C.: U.S. EPA.

―――. 1993b. *Streamlined Approach for Settlements with De Minimis Waste Contributors Under CERCLA Section 122(g)(1)(A)*. (OSWER Directive #9834.7-1D). Washington, D.C.: U.S. EPA, Office of Solid Waste and Emergency Response.

PART V:
Natural Resource Damages

9

Liability for Natural Resource Injury: Beyond Tort

Richard B. Stewart

Recent debate about environmental regulation and liability reveals a remarkable reversal of positions. Business traditionally has insisted that environmental law and policy be driven by economic analysis of the costs and benefits of environmental regulation and liability rules. Most environmental advocates and many government regulators have voiced strong opposition, asserting that a healthy and unspoiled environment is beyond price.

Recently, however, federal and state government agencies, with strong backing from environmental groups, have aggressively and successfully sought to impose statutory damage liability for injuries to public natural resources. In doing so, they have sought to base damage claims on contingent valuation methodology (CVM) surveys that seek to measure the economic nonuse value that individuals place on the environment by asking them how much they would pay to preserve a given natural resource from injury. Business, on the other hand, has attacked use of this new methodology as unreliable and unsound.

This novel debate over the role of economic methodology in environmental protection has been generated by recent statutes authorizing natural resource damage liability. In the 1980 Comprehensive Environmental Response, Compensation, and Liability Act (CERCLA), Congress authorized federal, state, and tribal trustee authorities to bring court actions to recover damages for injury caused by releases of hazardous substances to natural resources "belonging to, managed by, held in trust by, appertaining to, or otherwise controlled by" the trustees [CERCLA §§ 101(16), 107(a)(4)(C)]. Trustees may also recover the costs that they have incurred in assessing these natural resource damages (NRD). The Clean Water Act (CWA) and the 1990 Oil Pollution Act

(OPA) provide similar NRD liability for injuries caused by oil spills [CWA §311(f)(4), (5); OPA §1002(a), (b)(2)]. Recoveries must be spent to restore, replace, or acquire the equivalent of the injured resource; federal trustees are authorized to make such expenditures without further authorization or appropriation by Congress. These statutes impose NRD liability above and beyond the costs of removing the pollution in question. Many states have enacted similar statutes authorizing state officials to bring actions for natural resource damage.

To date, the most significant natural resource damage case is the $900 million settlement of NRD claims, pursuant to the Clean Water Act, by the United States and Alaska against Exxon for the Exxon *Valdez* spill in Prince William Sound. Government NRD recoveries in the tens of millions of dollars have also been obtained for hazardous releases or oil spills at other sites. The federal government alone has obtained NRD recoveries in respect of nearly fifty sites. Many additional claims are pending. Natural resource damage liability is becoming an important and widely used tool in government's armory of environmental remedies.

Problems in Applying Private Tort Liability Principles to Public Natural Resources

The NRD regime represents an extension of traditional tort liability to public natural resources and is based upon a simple and appealing logic. Private owners obtain compensation and redress for injury to their property. Property damage liability also serves important deterrent functions. The public commons deserves at least as much protection as does private property.

The effort to protect the commons through statutory extension of tort liability has, however, created a number of novel, difficult, and unresolved legal issues, including the standards for determining injury and causation and the measure of damages, the availability of jury trial, and the principles governing court review of trustee damage assessments. Resolving issues of injury, causation, and damages also involves extraordinarily difficult and contentious factual issues at the frontiers of science. The validity and reliability of contingent valuation methodology in assessing nonuse values is sharply disputed. These difficulties are merely symptomatic of a more fundamental problem, stemming from the hybrid character of the NRD scheme and the flaws inherent in the attempt to adapt private tort liability to public natural resource injury.

The NRD statutory programs represent a novel hybrid, composed of elements of tort, trust, and administrative law. In imposing tort liability for public natural resource injury, Congress went far beyond the common law by adopting rules of near-absolute strict, retroactive, joint-and-several liability against a wide range of parties. Another significant departure from the common law is the rejection of market measures of damages; instead, damage claims have been based on restoration costs, lost use values, and nonuse values as measured by contingent valuation methodology.

These novel damage measures reflect the public trust element in the NRD statutes. Congress empowered government agencies with management jurisdiction over natural resources to act as "trustees" to assess and recover damages. In order to preserve the public resource trust, the NRD statutes require trustees to spend recoveries solely for restoration of the injured resource.

Finally, the NRD statutes incorporate elements of an administrative model by empowering trustee agencies to make damage assessments and carry out restoration activities. The shield of a "rebuttable presumption" is lent to damage assessments conducted by trustees "in accordance with" NRD assessment regulations issued under CERCLA and CWA by the Department of Interior (DOI) and under OPA by the National Oceanic and Atmospheric Administration (NOAA). These regulations are required to "identify the best available procedures" for damages assessment [CERCLA §§107(f)(2)(C), 301(c); OPA §1006(e)].

The potential advantages of administrative efficiency and expertise, however, are undermined by the tort and public trust elements in the statutory design; the first insists on a civil jury trial, while the latter restricts managerial discretion by requiring that recoveries be spent exclusively to restore, replace, or acquire the equivalent of the injured resource. The various hybrid elements of the NRD system thus tend to contradict rather than complement one another, creating serious legal uncertainties and managerial and other institutional difficulties. By contrast, the cleanup provisions of CERCLA are aimed at traditional regulatory goals—protection of health and the environment—under the direction of a single federal agency, the U.S. Environmental Protection Agency (EPA).

The Perils of Pricing Nature. The effort to adapt the private law tort model to deal with injury to public natural resources, particularly those with special aesthetic, ecological, or historical value, is an inherently flawed enterprise. Environmentally significant resources are publicly owned because they have special values, including nonuse preservation values, that would not be protected adequately under a market-based system of private ownership. These special values will not be protected adequately by tort remedies for property injury, which are geared to market measures of damages.

Trustee agencies have sought to meet the problem of adapting private tort damages rules to injury to publicly owned natural resources by making replication of the pre-injury resource the presumptive remedy and by seeking additional recoveries based on the reduced value of the resource pending restoration. They have also invoked CVM surveys to measure the nonuse values placed by individuals on preserving a resource. These efforts, however, have created more problems than they solve.

Restoration of a resource to its pre-injury condition is often infeasible, excessively costly, and a waste of resources that could be better used elsewhere. Contingent valuation methodology is an unreliable and inappropriate means of measuring damages. The values obtained by CVM surveys are often grossly inflated and otherwise inconsistent with economic behavior. There is at least grave doubt whether contingent valuation methodology ever can measure reliably the economic values that individuals may place on the preservation of natural resources. Moreover, the values of environmental preservation are in large part noneconomic in character, based on society's mutual stake in safeguarding a common environmental heritage. It is therefore a mistake to treat the preservation of nature as a commodity. Accordingly, environmentalists who seek to use CVM surveys as a means to impose large natural resource damage liabilities are shortsightedly selling their birthright.

The Need for a Fundamental Rethinking of the NRD System. In large part because of its novel hybrid character and the effort to adapt private tort liability to public natural resources, the current NRD system involves large transaction costs, ubiquitous factual and legal uncertainty, bureaucratic waste, and the persistent threat of arbitrary and grossly inflated liabilities. These difficulties cannot be solved without fundamental rethinking and restructuring of the current system. Natural resources should be protected by means other than the private tort model. These means include criminal sanctions and civil penalties, trust funds generated by imposing taxes or fees on risk-creating activities or scheduled liability assessments following the model of workers' compensation.

DIFFICULTIES IN CURRENT NRD ASSESSMENT

This section of the chapter describes a range of conceptual and practical difficulties encountered in the administration of the current NRD system. Subsequent sections explain how many of these problems stem from the effort to apply the private tort model to the nonmarket values secured by public natural resources and from the overlay of public trust and administrative elements.

The NRD statutes authorize natural resource trustees to assess damages, obtain recovery from the parties responsible for the injury, and expend the recoveries to restore the injured resource. Damage assessment requires trustees to undertake a three-stage analysis: first, to determine the existence and extent of injury to the natural resource; second, to select the appropriate restoration remedy; and third, to determine damages, which include the costs of restoration, interim impairment of resource values, and the costs incurred by trustees in assessing damages [CERCLA §107(a)(1)(C); OPA §1006(d)(1)(C)]. Each step in this process is beset by conceptual, scientific, and policy problems.

Determining Injury and Causation

Following the tort model, liability is imposed for injury to natural resources caused by pollution for which the defendant is responsible. A trustee must, therefore, chart the effects of the pollution over time from the baseline of the resource's uninjured state. In doing so, a trustee typically is faced with a severe lack of existing data about the components, functioning, and prior history of the affected ecosystems. Accordingly, trustees must generally attempt to reconstruct a baseline, drawing on biology, ecology, toxicology, hydrology, and geology.

In determining the pre-injury baseline, a trustee often confronts a moving target. Even in environments that are not subject to significant human influence, populations of relevant species normally exhibit substantial variation due to population dynamics, intra-ecosystem variables, and exogenous factors such as climate (See Botkin 1990; Cross 1993; Stevens 1991). A decline, for example, in a fishery following a spill or a release may be due to natural population cycles. Even if a change is caused by human activity, it may not necessarily represent compensable injury. For example, contamination may kill individual fish, but the population may be unaffected because biological mechanisms of compensation may enable populations to maintain themselves despite individual losses. Moreover, the extent of any injury may not become clear for many years. Nature's capacity to cleanse and restore itself may allow recovery without human assistance. On the other hand, contamination may cause subclinical changes in organisms that result in manifest injury only years later. Finally, even if injury can be attributed to human activity, was it caused by a particular defendant? What liability does a ship responsible for a moderate spill in an already highly polluted harbor bear for shellfish contamination?

Case-by-case resolution of these factual and scientific issues, many of which lie at the frontiers of several branches of science and all of which require enormous amounts of data, entails enormous difficulty for trustees, high expense for potentially responsible parties (PRPs) who ultimately must foot the assessment cost bill and also conduct rival assessments, and potentially grave inaccuracy. The injury determination phase is, however, only the beginning of the damage assessment problem.

Choosing among Restoration Options

The next dilemma in damage assessment concerns the actions that the trustee should take to restore the injured environment. In *State of Ohio v. Department of Interior* [880 F.2d 432, 44–59 (D.C. Cir. 1989)], the court, in reviewing DOI's NRD assessment regulations, held that restoration cost is the presumptively correct measure of damages. Trustees must opt for restoration unless its cost would be "wholly disproportionate" to the diminution in value of the resources, in which case trustees should elect to claim for the lost value of the injured resource (880 F.2d at 459). All of the NRD statutes require that recoveries (other than reimbursement of assessment costs) be spent exclusively for restoration. For these purposes, "restoration" includes replacement or acquisition of the equivalent of the injured resource as well as its physical replication.

A trustee should evaluate a range of measures that might be taken to repair, replace, or acquire the equivalent of the injured resource, considering the extent to which they will contribute to restoration goals and their costs. Trustees should also consider the option of doing nothing. In judging the benefits of restoration options, the trustee must take into account the contribution of cleanup operations undertaken by or under the supervision of EPA or the Coast Guard and state counterparts, as well as the restorative capacities of nature. These considerations often present large predictive uncertainties. Also, restoration of damaged environments is a fledgling art; thus, the cost and ultimate success of particular restoration measures is often highly uncertain.

Underlying choices among restoration options is a fundamental conceptual problem. Is the goal to replicate, insofar as possible, the physical and biological environment as it existed before a release or spill? Literal replication is impossible, but a trustee could aim to achieve the closest possible physical and biological equivalent. In many cases, however, replication is likely to be extremely costly. Alternatively, resources can be understood in economic terms as assets that provide a flow of services to humans. Depending on how broadly or narrowly the category of relevant services is defined, a services approach could offer trustees considerably more flexibility and imply less expense by making a wider range of restoration options available, from which the trustee could and arguably must pick the most cost-effective.

Consider, for example, a scenic stream on public lands contaminated by mine tailings. Assume that it would be enormously costly to remove all of the contamination and restore the stream to a condition that would support a natural trout fishery. Another stream on nearby private lands, somewhat different in character but affording comparable trout fishing opportunities, could be acquired for much less. Since restoration includes acquisition of the equivalent, defining the injured stream in terms of the recreational and aesthetic services that it provides to humans would enable trustees to remedy the injury by acquiring the nearby stream. The tort model suggests use of a services approach and also a duty by trustees to minimize damages. The emphasis on regulatory flexibility apparent in the broad statutory definition of restoration and the administrative model also argues for the services approach. Environmentalists, however, may invoke the strong public trust element in the statutory schemes to insist that nothing less than replication of the uninjured physical and biological environment will suffice.

Assuming that resources should be defined in terms of the services that they provide, what flexibility should trustees enjoy in replacing or acquiring the equivalent of resource services that have been impaired due to injury? Can they replace the contaminated trout stream with one a hundred miles away? With a pristine lake containing bass? The NRD statutes provide no guidance on how geographically and functionally proximate a replacement must be to the injured resource.

Resource Damages beyond Restoration Costs. While restoration is the preferred measure of damages, *Ohio* asserted that at some point the cost of restoration might be so great as to be "wholly disproportionate" to the value of the resource and therefore inappropriate (880 F.2d at 459). In order to decide whether this point has been exceeded, a trustee will have to determine the value of the resource and compare it to the cost of restoration. Accordingly, where restoration (including replacement or acquisition of the equivalent of the injured resource) is "wholly disproportionate" in cost to its value or is infeasible, damages should be based on long-term diminution in value. Even where restoration is undertaken, trustees claim recover for the temporary diminution in value in the interim between injury and full recovery. Regardless of the rationale for an NRD award, all such recoveries must by statute be spent on restoration, replacement, or acquisition of the equivalent of the injured resource. The NRD statutes reject the common law precedent which allows plaintiffs to spend recoveries as they please. In addition, trustees can recover and reimburse themselves for the reasonable costs of damage assessment.

Beyond CERCLA's prohibition of "double recovery" [§107(f)(1)], the legislation provides no clear guidance for the conduct and content of lost value assessments, stating only that DOI's damage assessment regulations, which must be adhered to in order to benefit from the rebuttable presumption under CERCLA and CWA, must "take into consideration factors including, but not limited to, replacement value, use value, and the ability of the ecosystem or resource to recover" [CERCLA §301(c)(2)]. Following the tort model, the NRD assessment regulations promulgated by DOI in 1986 measured diminution in resource value based on the market price or appraisal value of the resource. If no such market-based measures existed, diminution in value was to be based on impaired use values, determined by factor income, travel cost, hedonic pricing, or contingent valuation methodology. Only in the event that neither market nor other measures of use value existed could diminution in value be based on impaired nonuse values, determined through contingent valuation methodology.

This hierarchy was rejected by the D.C. Circuit Court of Appeals as an unreasonable interpretation of CERCLA. The court found that it was too rigid and tended to neglect nonmarket and particularly nonuse values, which the court referred to as "passive use" values. Current NRD regulations and proposed regulations eschew a hierarchical structure and allow trustees broad authority to recover for use and nonuse values, subject to the prohibition against double counting. Also, trustees are free to use a variety of methodologies for measuring impaired use values, including market price, appraisal techniques, travel cost, hedonic pricing, and contingent valuation methodology. The latter is the only methodology recognized for assessing nonuse values. The problems in attempting to place an economic value on the distinctive nonuse components of environmentally significant public resources are severe, as explained below. But unless these regulations, too, are judicially invalidated, they will provide trustees with broad discretion and simultaneously with a claim for a presumption of accuracy for damage assessments. At the same time, they will leave trustees virtually rudderless and create the risk of large and unpredictable liabilities for PRPs. Trials will be battles between regiments of experts over mountains of factual material and mind-boggling methodological disputes.

LEGAL AND INSTITUTIONAL PROBLEMS CREATED BY THE CURRENT NRD SYSTEM

The hybrid nature of the current NRD system compounds the inherent difficulties of case-by-case assessment of damage liability for injury to public resources by creating significant legal uncertainties and institutional problems that make resolution of natural resource damage claims costly and time-consuming and foster resource misallocation and waste.

Legal Uncertainties

A central, unresolved legal uncertainty created by the hybrid NRD system is whether the right to jury trial attaches to NRD liability claims. Invoking consistent precedent that rejects jury trial claims in government cleanup liability actions under CERCLA, the government has argued that NRD claims are likewise not subject to jury trial because they are equitable and administrative in character and seek restitution of the injured resource. But two out of the three lower federal courts that have considered the issue have held that NRD claims under CERCLA are akin to tort actions and are therefore subject to jury trial.[1] Seventh Amendment precedent does not clearly resolve the issue, but the better view favors a right to jury trial, at least for recoveries based on lost value as opposed to restoration cost. Jury trial of injury, causation, and natural resource values will further increase cost and delay. If CVM studies are allowed into evidence, the likely variations between the estimates of nonuse value offered by the government and by defendants will be enormous. The jurors would have equally enormous discretion to award damages that could, at the upper range, amount to hundreds of million or billions of dollars.

Another open issue is whether trial of damage assessment issues should follow an administrative law model or a common law civil litigation model. The government has asserted that the evidence in an NRD action generally should be limited to the administrative record produced by trustee NRD assessments and that damage assessments should be reviewed judicially under the arbitrary and capricious standard applicable to review of administrative proceedings. In response, PRPs maintain that NRD actions are tort actions and that defendants sued for damages are entitled to a full, de novo trial. It is also difficult to reconcile on the administrative record with the right to jury trial.

The standard for judicial review of NRD settlements is also unsettled. Some courts, stressing administrative efficiency, defer to settlement reached by the trustees while others, stressing public trust principles, exercise close scrutiny. For example, in *Acushnet & New Bedford Harbor Trustees: Proceedings re Alleged PCB Pollution* [712 F. Sup. 1019 (D Mass. 1989)], the court approved the government's settlement of a PRP's NRD liability, emphasizing that settlement of NRD cases should be favored because disputes regarding the extent of damage, causation, apportionment, and unsettled legal questions would otherwise threaten huge litigation costs and lengthy delay. In contrast, in *State of Utah v.*

Kennecott Corp. [801 F. Supp. 553 (D. Utah 1992); *appeal dismissed*, 14 F.3d 1489 (10th Cir. 1994)], the court overturned the state's settlement of NRD claims for pollution of an aquifer, applying a markedly less deferential standard and faulting the settlement for, among other defects, failure specifically to earmark a recovery for lost nonuse values.

A further unresolved legal puzzle is created by the statutory provisions that accord a rebuttable presumption of correctness to trustee damage assessments made in conformity with DOI or NOAA damage assessment regulations. An initial question is whether the NRD assessment regulations issued by DOI [59 *Federal Register* 14262, March 25, 1994] and proposed by NOAA, which leave trustees enormous case-by-case discretion, provide sufficient direction and assurances of reliability to trigger the presumption. A second question is the effect of such a presumption, if applicable, and its rebuttable character. Does the presumption merely relieve a trustee of the burden of coming forward with evidence, or does it give it some more powerful advantage? Also, an evidentiary presumption is a civil litigation concept difficult to reconcile with the administrative model of judicial review advocated by trustees.

Institutional Problems

In addition to the legal uncertainties noted above, three institutional features of the current NRD regime lead to significant waste and delay: multiple parties, lack of coordination between restoration and cleanup activities, and the statutory limitations on expenditure of NRD recoveries.

The sweeping net of liability created by CERCLA leads to the presence of multiple defendants in NRD actions. The problems thus created have emerged clearly in CERCLA cleanup litigation: enormous transaction costs, delays in undertaking remedies, and difficulties in reaching settlement. The NRD system exacerbates such problems by multiplying the number of plaintiffs. Congress gave cleanup authority to EPA (in the case of hazardous substance releases under CERCLA) and the Coast Guard (in the case of oil spills under CWA and OPA). But Congress gave NRD authority to a large number of separate trustee agencies (federal, state, and tribal) with overlapping jurisdictions and no predetermined hierarchy or allocation of authority among them.

First, problems are created when more than one trustee agency asserts authority over a given resource. The seemingly straightforward trust concept in practice fits awkwardly with the complex overlay of ownership, regulatory, and management authorities exercised by a variety of federal and state agencies as well as Indian tribes. In the Exxon *Valdez* spill, for example, no fewer than six federal and state agencies asserted trustee authority. This proliferation of plaintiffs leads to wasteful turf disputes that impede coordination of damage assessments, make restoration decisions and settlements more difficult, and delay restoration. The statutes give no criteria to resolve such conflicts.

Second, the division of remediation authority between cleanup agencies (such as EPA and the Coast Guard) and trustee agencies responsible for restoration creates additional problems of coordination and consistency. Cleanup remedies imposed by EPA in some cases have caused additional natural resource injury or have conflicted with restoration measures. On the other hand, state trustees are beginning to use the threat of NRD liability to require fuller cleanups than required by CERCLA. CERCLA's goal of protecting health and the environment, for example, can often be achieved by less than complete removal of contamination. State trustees, however, have used the threat to require complete removal as a restoration measure in order to inflate NRD settlements.

Finally, the NRD statutes depart from normal budgeting and spending rules by requiring that all damage liability recoveries be spent exclusively to restore, replace, or acquire the equivalent of injured resources or to reimburse damage assessment costs incurred by trustees. Moreover, federal trustees are authorized to expend recoveries for these purposes without further authorization or appropriation by Congress. These provisions, which reflect a public trust rather than the tort model, produce several problems.

Trustees may recover only those damage assessment costs incurred in response to particular spills or releases. This discourages trustees from conducting baseline surveys and other assessment work prior to a given spill or release, even though such work may be more cost-effective and scientifically sound as well as less adversarial than damage assessments conducted after a spill or release. Moreover, trustees assign their own, already budgeted personnel to do NRD assessments in particular cases and to seek to recover the costs involved, plus a general indirect overhead charge, without further budget control. These arrangements encourage excessive trustee NRD assessment efforts.[2]

The requirement that damage recoveries be spent exclusively for restoring, replacing, or acquiring the equivalent of the injured resource creates further inefficiency and waste. If restoration is defined in terms of physical replication, this requirement imposes remedial inflexibility that ill serves the public interest. Many injured public resources are not rare or exceptionally scenic or otherwise very valuable, yet the costs of replicating their pre-injury condition may be extremely large. Even if it were accepted that recoveries should be earmarked for environmental purposes, the public would in many cases be far better served by spending at least a part of recoveries to enhance or expand other public natural resources.

An approach to restoration based on resource services could afford trustees greater flexibility, but in the absence of the safeguards of control and accountability provided by the normal budget process, that flexibility could be abused. The lack of these safeguards, combined with the absence of firm requirements in the NRD assessment regulations that restoration efforts be reasonable in relation to benefits and cost-effective, creates an incentive for trustee agencies to "gold plate" restoration activities, loading on overhead charges and designating their own employees to undertake damage assessments and restoration activities in order to maximize recovery of off-budget moneys and expand their operations. A large part of the enormous Exxon *Valdez* damages thus far paid by Exxon have been spent by the six federal and Alaska trustee agencies to reimburse their own expenses rather than to restore Prince William Sound. (U.S. GAO 1993; Salmon Fishermen 1993) The responsible "polluter pays" in the first instance, but ultimately society as a whole, foots the bill for waste and misdirected resource outlays.

THE PRIVATE TORT MODEL: FUNDAMENTAL FLAW IN THE CURRENT NRD SYSTEM

While the current NRD system is a hybrid, the dominant element is tort. The tort element is responsible for many of the problems of the NRD system previously discussed, including the difficulties of case-by-case determination of highly complex environmental injury and causation issues and an array of legal uncertainties. Tort measures of damages are not suitable for the noneconomic values inherent in public natural resources. CVM mistakenly attempts to commodify these values. The unreliability of CVM surveys adds enormous unpredictability and controversy to the damage assessment process.

Private Tort Liability's Inapplicability to Public Natural Resources

Two dominant rationales support the imposition of tort liability when one person injures another's person or property: corrective justice and welfare maximization. Neither justifies extending tort liability principles developed in the context of harm to private property to injury to public natural resources.

Corrective justice understands the function of tort liability as redressing the defendant's violation of the plaintiff's rights. Violation by one person of another's rights disrupts the moral equilibrium. Redress is obtained by awarding damages to restore, insofar as the law can, the preexisting equilibrium. These principles do not justify a damage award in favor of the government against one who has injured the government's property. The government is a political and legal abstraction, not a morally autonomous person with rights. As such, it has no claim to redress of its own rights to liberty and security. Its injury is impairment of its ability to serve the public. Moreover, the supposition that the defendant is a wrongdoer is undercut severely by the fact that NRD liability is near-absolute. Principles of distributional equity or efficient resource use, as opposed to corrective justice, may dictate that those whose activity has caused injury to public resources, rather than taxpayers in general, finance restoration of the public services loss as a result. Tort liability, however, is only one among several possible means of providing the necessary resources. Other means include taxes on risk-creating activity, civil penalties for regulatory violations, and administrative assessments of scheduled damages similar to workers' compensation. The choice among these mechanisms is not dictated by corrective justice, but by considerations of administrative cost and practicality, distributional equity, and the importance attached to incentive functions. As developed below, these considerations favor alternatives to the use of tort liability to deal with injury to publicly owned natural resources.[3]

From a utilitarian or welfare economic perspective, the purpose of liability rules is to promote aggregate societal welfare. They may do so by advancing two distinct goals: economically efficient accident reduction and reduction of the costs of bearing the risk of accident losses. In the calculus of aggregate welfare, legal instruments for achieving these two goals must also be evaluated in terms of their administrative and error costs.

In theory, a rule of strict damage liability provides appropriate incentives for caretaking and also ensures that industries engaged in handling oil or chemical wastes bear the costs of the injuries associated with their activities. In practice, these functions are impaired by serious problems in implementing the NRD liability system. The great scientific uncertainties in determining injury and severe difficulties in measuring the economic value of public natural resources means that NRD assessments are likely to be highly inaccurate, resulting in inappropriate incentives. Also, PRPs may avoid full liability by using superior resources to wear down the government in the costly litigation entailed by the NRD liability regime. On the other hand, the threat of uncertain and potentially enormous NRD liabilities creates a significant danger of overdeterrence. There is no assurance that these various errors will cancel out neatly. In practice, NRD liability will not create appropriate caretaking incentives.

As for risk bearing, the government is a better risk bearer than any private party because of its size, wealth, and the diversified character of its activities. Considerations of risk bearing accordingly do not justify imposition of tort liability for natural resource damages on private parties. Finally, the administrative costs of the NRD liability system are extraordinarily high. Tort liability for injury to public natural resources is thus very unlikely to be the best means for promoting societal welfare. As developed below, better tools are available.

Noneconomic Values of Environmentally Significant Public Natural Resources

Environmentally significant public natural resources have important nonuse values that are primarily noneconomic in character. This circumstance helps explain some of the severe problems encountered in developing a reliable economic measure for such values and also explains why the effort to commodify such values is fundamentally misguided.

The benefits from using a resource may either be excludable or nonexcludable. They are excludable if use of the resource can be limited to persons who pay for the benefits in question, while those who do not pay can be selectively excluded. A resource must have the characteristic of excludability in order to create private property rights in the resource. Many of the use benefits associated with land or delineated bodies of water are excludable. There are, however, some types and uses of resources, such as airsheds, that do not sufficiently satisfy the criterion of excludability to make private ownership feasible or appropriate. The benefits of cleaner air accrue to everyone within the airshed. Because it is not feasible to provide cleaner air only to those who pay a fee and exclude those who do not, it is not possible to establish a private market in clean air. Cleaner air can only be established through collective government action, such as regulation, a system of transferable pollution allowances, or emission fees.

Certain important public resources are characterized by nonexcludability, including the atmosphere and the territorial sea. But many natural resources owned or intensively regulated by government, including the public lands, are characterized by physical excludability. Given the traditional presumption in favor of private property in the United States, what justifies continued government ownership of such resources, particularly those that are environmentally significant? This question can best be answered by considering the different use and nonuse values associated with environmental resources.

Most environmentally important natural resources have significant use values—such as viewing, hiking, camping, fishing, and boating—

associated with pristine environments, scenic vistas, and rich, diverse, or unusual ecosystems. The special qualities that make such resources unusually beautiful or significant often require unified management of large tracts of land and water. This task is unlikely to be accomplished adequately by the market. The assembly and maintenance of such tracts through private ownership would require enormous fixed capital investments and would also be impeded by holdout problems and high transaction costs. Government potentially can correct this market failure by acquiring, controlling, and managing these large tracts in order to secure the special use benefits that they support. Intensive regulation of private lands is an alternative possibility, but such regulation would have to be so comprehensive and intrusive as virtually to amount to government ownership.

The value of environmentally significant areas, however, extends beyond an individual's current or future use. For example, there may be pure existence values associated with environmentally significant resources. Individuals may take satisfaction from the assurance that rare, beautiful, or ecologically significant environmental resources will be preserved, regardless of any potential use that might be made of them by anyone. These nonuse benefits of preservation are not, however, excludable. An entrepreneur who owns a habitat occupied by endangered species has no way of charging those who derive "long-distance" nonuse benefits from preserving the species, or of denying such benefits to those who fail to contribute to the preservation effort. Accordingly, a market-based system of private property will fail to provide a sufficient level of nonuse benefits. Government ownership and management potentially can correct this failure.

While it is thus possible to characterize nonuse values in the economic model of preference satisfaction, the commitment to preserve natural resources generally reflects a broader array of collective concerns. Individuals support nature's protection for a variety of noneconomic, altruistic reasons (Singer 1975; Taylor 1986). Many strongly support preservation of nonhuman species and natural ecosystems on the grounds that humans owe ethical duties to nature or otherwise have obligations to maintain the integrity of the ecological community of which they are a part. Many also support preservation for the benefit of other humans, including future generations. Members of a political community justifiably may seek to assure access by all to diverse experiences and associated values, including those that are not provided, or not provided adequately, by a market economy. Such "noncommodity" values are not necessarily superior, but, as J.S. Mill argued, they are an essential element in the education and self-development provided by diverse, reflective experience. The market alone will not provide an adequately rich and diverse set of the experiential opportunities that should be available in order to facilitate each individual's search for the good life. Government, in combination with nonprofit and voluntary organizations, can help provide that assurance (Stewart 1983). The opportunities to visit or merely to learn about and appreciate rare, beautiful, or ecologically significant sites, species, and ecosystems are among those that should be ensured collectively.

Members of a political community also support preservation of exceptionally scenic or historic resources because they are symbols or expressions of that community. For Americans, great scenic areas such as Yellowstone, the Grand Canyon, and the undeveloped regions of Alaska have acquired such a status. Europe has its cathedrals and castles; the United States, its natural scenic wonders. In visiting such sites, or simply knowing that they are secured as part of the common national heritage, people learn and reaffirm their identity as members of a political and social community that maintains its identity over time. Accordingly, preservation of those sites reflects a joint commitment to maintaining a common heritage.

Many of these collective, noneconomic rationales for resource preservation have a democratic character: all members of society should have access to a diverse range of experiences, including the experience of rare, beautiful, or wild environments, and all Americans have a common share in certain sites that are part of the national heritage.

The Distortion of Nonuse Values by Economic Measurements

As previously explained, the nonuse benefits of resource preservation lack the characteristic of excludability. Accordingly, the nonuse values of public natural resources cannot be measured by market prices for natural resources or by other use valuation methods, such as travel time or hedonic pricing, based on individuals' revealed preferences.[4] Trustees accordingly have sought to use contingent valuation methodology to determine nonuse values. This effort has provoked enormous controversy. CVM surveys have yet to meet minimum standards of reliability. This failure is due to the inherent difficulty in creating a realistic but necessarily hypothetical market for nonuse values and to the fact that nonuse values are in substantial part noneconomic.

The CVM issue was framed by the D.C. Circuit in *Ohio*, which interpreted the NRD legislation to require, in principle, that "passive use" values, such as option and existence values, be included in the calculation of lost value, while also requiring that values for which trustees claim compensation must be "reliably calculated," in accordance with

CERCLA's mandate that DOI regulations prescribe "best available procedures" for damage assessment (880 F.2d at 464, 476–77).[5] Applying these standards, the court upheld DOI's authorization of contingent valuation methodology on the basis of studies of the reliability of the methodology available to DOI as of 1986, most of which concerned use values with which survey respondents were familiar rather than nonuse values. DOI's conclusion—that contingent valuation methodology was an appropriate assessment methodology in at least some cases, subject to qualitative requirements such as the use of "sophisticated questioning" to promote greater accuracy and an overall assurance that assessments be reliable—was sufficiently reasonable to merit judicial deference. Accordingly, the court rejected a facial challenge to the administrative determination that contingent valuation methodology may be used in assessing "passive use" values (880 F.2d at 475–478).

Since the *Ohio* decision, both the volume of CVM studies of nonuse values and debate among economists over the reliability of contingent valuation methodology have grown enormously (Mitchell and Carson 1989; Kopp and Smith 1993; Kahneman and Knetsch 1989; Smith 1992; Kahneman and Knetsch 1992b; Rowe and others 1991; Rubin, Helfand, and Loomis 1992; Dobbins 1994; NOAA 1993; Freeman 1993; Note 1992). Many recent studies and analyses have identified previously unrecognized problems with contingent valuation methodology. Recent research also shows that previously acknowledged drawbacks of this methodology are more serious than originally was believed. CVM researchers have thus far been unable to correct many of these defects.

A fundamental challenge for CVM surveys is the difficulty in creating a realistic hypothetical market in which the nonuse "commodity" in question is described accurately and respondents are induced to make the choice that they would make if they were spending real money and operating under a real budget constraint. It is very difficult to devise a survey instrument that will describe concisely and accurately the nonuse aspect of a resource and its impairment in terms intelligible to the average person, particularly given scientific uncertainty and the need to factor in the positive effects of restoration activities in determining net injury. It is also difficult to establish meaningful budget constraints and a realistic payment vehicle that excludes strategic behavior in order to simulate a real purchase decision. Survey respondents often have difficulty grasping the basic concept of nonuse value, and individuals are unlikely to have well-developed monetizable preferences for nonuse resource values that are not and that cannot be traded in markets. Moreover, the survey must deal with the fact that answers are often colored by general attitudes—such as overall support for environmental protection or antipathy to corporate polluters—rather than representing a specific evaluation of the particular resource in question.

Recent studies have shown that CVM results for nonuse values are extremely sensitive to the elicitation format and to the sequence and wording of the questions (McFadden and Leonard 1993; Kahneman 1986), and that preferences for one commodity over another can even be reversed depending on the format (see Kahneman 1986 and studies cited therein). While survey designers claim that these problems can be overcome, the persistence of large variations in responses produced by modest variations in the framing of survey questions indicates that the problems may be serious and persistent. Unresolved controversy remains over how to define the relevant population to be surveyed and how to deal with respondents who were previously unaware of the resource in question.

Moreover, the results of CVM studies of nonuse values are inconsistent with standard economic behavior and principles of rational choice, in the following respects:

- *Indifference to quantity*: Respondents' valuations do not vary appropriately with the amount of the resource being offered (Kahneman and Knetsch 1992a; Desvousges and others 1993). For example, in one study respondents stated that they would pay the same amount to save 20,000 birds and 2 million birds.
- *Embedding*: The valuation of a resource is affected significantly by whether it is offered alone or in connection with other resources (Kahneman and Knetsch 1992a; Kemp and Maxwell 1993, Cummings and Harrison 1992). For example, if asked how much they would pay to preserve an endangered species, such as the northern spotted owl, respondents may state that they would pay $75 or $100. But if presented with a larger number of species and other natural resources and asked how much they would pay to preserve each, the stated willingness-to-pay for the owls drops significantly.
- *Sequence*: Valuation is affected significantly by the location of a resource in a sequence with other resources.
- *WTP vs WTA*: Very large disparities exist between willingness-to-pay (WTP) and willingness-to-accept (WTA) measures of economic value, disparities that cannot be explained by income effects created by the assignment of the entitlement to the resource to the person who injures it (WTP) or to those who wish to preserve it (WTA).[6]
- *Bimodal responses*: Survey answers are characterized by many zeroes and many very high responses.
- *Inconsistency with actual expenditure decisions*: Hypothetical CVM values for public environmental goods do not reasonably approximate the amounts that people actually contribute to those goods

(Seip and Strand 1991; Duffield and Patterson 1992; Desvousges and others 1993).
- *Implausibly high responses*, such as an asserted WTP of $32 billion *annually* to save the whooping crane. By contrast, annual giving to all environmental nonprofit organizations in the United States in 1991 amounted to $2.5 billion (Mead 1993).

In connection with its development of NRD assessment regulations under OPA, NOAA convened a distinguished panel of economists chaired by Professors Kenneth Arrow and Robert Solow. The panel's report (58 *Federal Register* 4601, 1983) confirmed these difficulties and set forth criteria that a CVM study of nonuse values must meet in order to generate results that might be considered reliable.[7] As the panel acknowledged, no existing CVM study satisfies these criteria. There remains a substantial question whether the methodology can be developed to the point where it is capable of doing so. The Environmental Economics Advisory Committee of EPA's Science Advisory Board recently found that a "state of the art" CVM study of the nonuse value of groundwater (McClelland and others 1992) was seriously flawed.

The NOAA panel acknowledged the shortcomings of CVM, set forth criteria to address them, and indicated that contingent valuation methodology might be improved to the point where CVM surveys could provide a "starting point" for damages determinations by judges and juries. The panel conceded that even state-of-the-art nonuse CVM valuations tend to produce greatly overstated estimates of nonuse values. The panel would deal with this problem by having judges and juries decide the extent to which survey results must be scaled in order to reflect "true" economic preferences (58 *Federal Register* 4610). It is, however, bootless to suppose that the failure of CVM researchers to develop an objective measure of nonuse values can be cured by turning the problem over to the unguided intuitions of judges and jurors. In such a situation, revision through litigation of an already arbitrary number would be tantamount to plucking a damages figure out the air. No court has held that contingent valuation methodology is a reliable and appropriate method for valuing nonuse damages.[8]

Recent psychological research has concluded that responses to nonuse CVM surveys do not reflect economic valuation of the particular resource under study, but rather generalized attitudes and feelings about the environment, ethical values about injury, moral satisfaction obtained by supporting a "good cause," symbolic statements of the importance of the environment, perceptions of civic duties, or an informal, untutored social cost-benefit analysis (Schkade and Payne 1993; Kahneman 1986). These studies are consistent with the noneconomic justifications for environmental preservation discussed earlier and help explain why the results of CVM surveys are not consistent with observed economic behavior.

Commodification of Environmental Values Does Not Justify the Full Compensation Argument

Environmentalists have argued that contingent valuation methodology, despite its limitations, must be used in NRD assessments, for otherwise trustees will not recover for impaired nonuse values, the public will not be made whole for the injury suffered, and polluters will not be deterred adequately. This argument is flawed in several important respects.

The "full compensation" argument is based on corrective justice conceptions that are, as discussed previously, inappropriate in the case of public natural resources. But even if a private tort model of liability were applicable, it would not support an award of damages based on the speculative results of CVM surveys. Even in private tort actions, courts do not permit damages for losses unless they can be determined in a reliable way without excessive cost. For example, the law generally limits recoveries for pain and suffering or emotional distress to the physically injured victim and, in some jurisdictions, members of the victim's immediate family. Many others plausibly may claim psychic loss by reason of the victim's injury, including friends or even strangers who learn of the victim's plight through the media. The law nevertheless denies recovery for such losses[9] for two important reasons.

First, the law insists on certain minimum requirements of accuracy and reliability in the determination of damages. Damage measures that are unreliable and unduly subjective result in large variations in the damages awarded in similar cases, producing results that justifiably are regarded as arbitrary and unfair. Second, the law is concerned to avoid the large transaction costs involved in attempting to measure elusive elements of damages that might be claimed by large numbers of individuals. Both of these considerations militate strongly against efforts to base NRD on nonuse losses supposedly suffered by large populations of individuals, most of whom have never even visited the resource in question. The law does not insist on absolute precision in damage measures. But the inherent flaws of contingent valuation methodology, as well as the severe problems in applying it, make it qualitatively different from the types of proof that the law generally allows in damage liability cases.

The most fundamental defect in the "full compensation" argument is the assumption that the nonuse value of public natural resources can be measured appropriately by aggregating individual economic preferences for the "commodity" of preservation. As already noted, many of the values that underlie preservation of environmentally significant resources are altruistic and collective in character rather than economic. The implicit suggestion of contingent valuation methodology—that the preservation of the Grand Canyon or the blue whale can be valued solely in terms of each individual's isolated self-interest, however refined—is wholly unconvincing. Survey responses are likely to reflect concern over nature's desecration, an altruistic commitment to preserving the environment as a common heritage, and a moral preference for a society in which environmental destruction would not occur. These values are highly relevant to the justification of preventive or remedial measures collectively undertaken, but not to the economic measures of loss appropriate in damage liability determination. Impermissible double-counting of economic loss arises if each individual considers others' welfare in addition to his own in making a valuation. In order to strip away the seemingly pervasive if implicit interdependencies and ethical elements in natural resource valuation, CVM surveys would have to use unprecedented surgery in order to isolate whatever residuum of value might be based solely on individual economic preference.

Accordingly, the argument that omission of nonuse values would lead to serious undercompensation of the public for injury rests on a category mistake. If nonuse values are principally noneconomic, then their omission from an economic measure of damages does not make the resulting damages inadequate. The common law flatly denies compensation for elements of loss based on compassion and ethical concern for others, such as the pity felt by the public about victims of widely publicized accidents. NRD law ensures full restoration of the injured resource in most cases, as well as compensation for interim lost use values. This is full compensation on the tort model.

Finally, recovery for nonuse values is not required in order to ensure adequate deterrence. As explained in the next section of this chapter, if regulatory requirements do not provide adequate caretaking incentives, the gap should be met by a system of civil penalties or scheduled damages rather than a liability lottery.

ALTERNATIVES TO THE PRESENT NRD SYSTEM

The defects of the existing NRD regime are so profound as to warrant fundamental rethinking of the rationale and the need for any case-by-case system for imposing liability on particular parties assertedly responsible for given injuries to public natural resources. Reconsideration points to the conclusion that the existing litigation model should be legislatively replaced by an administrative system of scheduled assessments based on the environmental risks posed by oil spills and hazardous substance releases, perhaps supplemented by taxes on risk-creating activities or by an expanded system of civil penalties.

The environmental risks posed by oil and hazardous substances currently are subject to comprehensive regulation. The Resource Conservation and Recovery Act (RCRA) establishes a stringent "cradle to grave" system of regulation for toxic wastes. A wide variety of regulatory programs, including those established by OPA, are aimed at promoting safety in the storage and transportation of petroleum and petroleum products. These regulatory requirements are enforced through a formidable array of criminal, civil, and administrative sanctions.

There are, nonetheless, limitations to the regulatory regime that may justify the adoption of additional measures. First, there are inevitable gaps in both the coverage and enforcement of regulations covering activities as ubiquitous and potentially hazardous as the handling of oil and chemical wastes. Second, a regulatory system may be faulted as failing to internalize to the regulated activity its full social costs because the activity bears no liability for the residual injuries that occur notwithstanding compliance with the regulations.[10] Third, distributional justice may require that private activities that threaten or cause harm to public natural resources bear the costs of restoring the public services that such resources provide.

Regulation usually is supplemented by tort liability, which provides general, "backup" incentives for caretaking. A system of strict liability may also serve cost-internalization and distributive justice goals. As previously explained, however, the special characteristics of public natural resources virtually disable the tort liability system from achieving these goals fairly and effectively in the NRD context. Other means of achieving these objectives must be canvassed.

One option is to impose a fee or tax on activities creating a risk of injury, set at levels to generate aggregate revenues roughly equal to the amounts necessary to redress the injuries caused by such risks. Such a system would provide a very rough form of cost-internalization and distributive justice. It would not, however, provide incentives for caretaking. Each firm engaging in the taxed activity would pay the same tax, regardless of whether it takes care to prevent injury or not.

The challenge is thus to devise a system that would provide caretaking incentives, as well as fulfilling the goals of cost-internalization and distributive justice, in a fair and cost-effective manner. These goals might best be secured by eliminating proof of injury or causation in given cases and imposing strict liability administratively for the release of oil and toxic substances into the environment, using a schedule of damages based on the average harm posed by the type of release in question. The closest analogy is the workers' compensation system, which not only standardizes the determination of loss earnings but typically deals with the nonmarket elements of personal injury by placing a fixed dollar value on life or limb through a schedule adopted by the legislature or an administrative agency. The schedule achieves only a crude justice, because it ignores many variations among individual cases. But the justice that it provides is far swifter and cheaper, and more predictable and consistent, than the individuated but costly processes of the tort liability system.

An NRD System of Scheduled Damages

Workers' compensation, however, still requires case-by-case proof of injury and an attenuated form of causation, elements that are unproblematic in most workers' compensation cases, but excruciatingly difficult in many NRD cases. In order to deal with the special characteristics of natural resource damages, what is needed is a system of scheduled damages based on the average injury caused by a spill or release of a given character in a given environment. Congress or an administrative agency could develop a schedule for assessments based on the volume and character of oil spilled or hazardous substances released and the quality of the receiving environment. The level of the assessment would be set to reflect, in a crude fashion, the average harm, measured in terms of the average cost of restoring the services to the public lost as a result of the resource injury, associated with different-sized spills in different locations. Because liability would be the average injury caused by a given type of release rather than the cost of restoring or replacing a specific resource, the liability of multiple contributors to spills or releases would be separate rather than joint. Following the workers' compensation model, these assessments should be exclusive of all other government civil recoveries for injury to public natural resources, although the government would remain free to impose civil or criminal penalties for violation of applicable regulatory requirements.

This approach would involve various problems of design and implementation. A major issue would be how far the schedule would attempt to "fine tune" assessments by reference to the toxicity of the spill or release in question and the nature and quality of the receiving environment. For example, for quantity, categories could be de minimis (0 to x pounds), small (x to y pounds), medium (y to z pounds), and so forth. A similar rough categorization of the toxicity of different families of chemicals could be developed. The quality and vulnerability of the environment into which the release occurred could also be similarly classified. These various classifications could then be combined in a matrix to yield a total numerical ranking that would be translated into an assessment of a given amount for a particular spill or release in a particular environment.

Certain forms of NRD liability provide a precedent for such a system. For example, an Alaska statute provides for a system of no-fault civil penalties based on the volume of oil spilled (Alaska Rev. Stat. Sec. 46.03.755). NOAA has also proposed several systems of simplified damage assessment under OPA (59 *Federal Register* 1062, 1118–1125, 1176–1177, 1994). This proposal, however, is legally vulnerable because it departs far from the tort model of case-by-case damage assessment embraced by OPA. DOI is also developing, under CERCLA, a simplified "Type A" damage assessment process for small releases of hazardous substances, based on the volume and toxicity of the release and the nature of the resource affected.

The NOAA and DOI proposals would, however, deal only with small, routine spills and releases. There is much greater need for scheduling in the cases of larger spills and releases because of the enormous transaction costs and large uncertainties involved in a case-by-case approach to injury determination and damage assessment. The factual decisions to be made in applying the schedule would be far fewer and simpler than those involved in the effort to determine damages from scratch in each case. There would be no jury. Determinations as to restoration or replacement of injured resources would take place in an entirely separate process.

A Restoration Trust Fund for a Scheduled NRD System

How should the moneys generated by such a schedule be spent? Under CERCLA, CWA, and OPA, damages are earmarked for restoration of the particular injured resource. This earmarking short-circuits the normal budget processes for determining and reexamining spending priorities. There is no good reason to suppose that spending all of the natural resource damages recovered in a given case to restore the particular resource injured is the socially best use of the recoveries. Moreover, under a scheduled approach, assessments would be based on the average harm caused by a given type of spill or release and accordingly in most cases would be either less or more than the amount actually needed to restore a given resource.

These considerations argue strongly for funneling all assessment revenues into the general Treasury account; they would be spent only as Congress directs through the normal budget process. The popular appeal of public trust notions and the political power of environmental groups may, however, dictate that assessments be paid into a revolving restoration trust fund. Initial financing of such a fund could be provided by tax on activities, such as oil transportation or hazardous waste generation, that threaten injury to natural resources. The fund would be replenished by scheduled assessments for oil spills or releases of hazardous substances.

Assessments would be earmarked for restoration purposes, but trustee agencies would not be required automatically to restore or replace each injured resource to its pre-injury level. Greater flexibility in the use of funds would reduce the current system's incentives to "gold plate" restoration projects and would enable trustees to provide greater environmental benefits to the public than would be possible if all recoveries were earmarked exclusively for restoration of the injured resource. Public trust concerns could be met by requiring trustees to explain, after opportunity for public input, decisions regarding use of assessments, including the level of restoration chosen for a given injured resource.

The restoration fund could provide moneys up front for restoration planning and work, removing the pressure on the government to settle cheaply in order to get started on restoration quickly. A scheduled system could also be designed to ameliorate jurisdictional conflicts that, under the current system, impede prompt resolution of claims and restoration as competing trustees haggle. When more than one trustee asserts a substantial interest in a resource, recoveries from a particular spill could be divided in accordance with some predetermined formula like those used to divide between the federal and state governments revenues from federal offshore oil or onshore mineral leases.

Enhancement of Civil Penalties

A supplement or alternative to scheduled assessments would be to modify the existing system for assessing civil penalties for violation of applicable statutory and regulatory requirements, including those in RCRA and OPA, to include threatened injury to public resources as an aggravating factor along with the economic benefit and offense gravity factors that EPA and other agencies already use to assess civil penalties for regulatory violations. Funding for restoration could be provided on a pooled basis by channeling penalty payments to a public resource protection fund.

Another alternative would be to merge the NRD regime into the cleanup programs. The overall goal would be to remove pollution and remedy natural resource injury through an integrated process, forsaking the tort model entirely. This approach would finesse the CVM debate and other disputes over lost resource values and target efforts on the overriding goals of protecting health and the environment and preserving the natural resource heritage. The revenues to achieve this integrated goal could be generated by a scheduled system of assessments for pollution discharges and releases. Such an approach would greatly reduce the transaction costs created by the current artificially separated systems of cleanup and restoration and their reliance on case-by-case liability determinations.

CONCLUSION

NRD liability is an experiment that must be reassessed in light of experience and reason. Basic changes in the current hybrid system are needed in order to develop a fair and efficient system for protecting public resources and raising the funds necessary for restoration. The most fundamental defect in the current system is the uncritical assumption that private law principles of tort liability should govern injury to public environmental resources. This category mistake must be corrected in order to rethink the existing NRD programs and develop a more just and effective system of remedies for protecting public natural resources.

ENDNOTES

[1] See *United States v. City of Seattle* [No. C90-395WD (W.D. Wash. Nov. 28, 1990)]; Acushnet River and New Bedford Harbor: Proceedings re Alleged PCB Contamination [712 F. Supp. 994, 1000 (D. Mass. 1989)]; and *United States v. Allied Chemical Co.* [No. C-83-5898 (N.D. Calif.; bench ruling)]. In all three cases, jury trial was demanded by the defendants. OPA damage claims based on spills in coastal or other waters subject to federal admiralty jurisdiction do not give rise to a jury right.

[2] The need to fund assessments out of current budgets, subject to uncertain reimbursement at some future point, imposes a countercheck. Some current legislative proposals would amend CERCLA to enable federal trustees to draw on the Superfund to fund damage assessments, thus removing this check.

[3] The circumstance that government agencies responsible for managing public resources may be regarded as trustees for the public does not warrant a different conclusion. The obligation of public trustees is to protect public resources; nothing in the trustee concept requires that any particular form of legal proceeding, such as a tort action, be used. Indeed, to the extent that alternatives to tort actions are more effective in protecting public resources, government officials would arguably be shirking their trustee responsibilities if they adopted the less efficacious remedy.

[4] Contribution to environmental and preservation causes is a form of revealed preference. However, the objects of such contributions generally are not sufficiently focused to provide a basis for measuring the nonuse value of a

given injured resource. Also, such contributions are tax-subsidized. On the other hand, free-rider problems may reduce the level of voluntary contributions below the aggregate willingness of individuals to pay for resource preservation.

[5] The requirement that any assessment method on which a trustee relies must be reliable to support recovery arises not only under CERCLA's specific regulatory mandate, but also under the federal rules for admissibility of evidence, and is axiomatic in torts suits. See Fed. R. Evid. 702, 703; and *Daubert v. Merrell Dow Pharms., Inc.* [113 S. Ct. 2786 (1993)].

[6] CVM researchers typically select the lower WTP measure rather than WTA, claiming that in doing so they are being conservative. But the enormous disparities between WTP and WTA tend to show that both numbers are arbitrary and unreliable. It is not sound, conservative practice to rely on one of two arbitrary numbers just because it is the lower.

[7] These criteria would require a study to accomplish five tasks: generate values that are consistent with the structure of revealed preferences of consumers in actual markets; account for variables in survey design and format that may elicit values that vary depending on the mode of eliciting valuations and account for outlier, nonresponse, and protest answers; ensure that respondents consider all relevant substitutes, observe budget constraints, and provide responses that reflect sensitivity to the magnitude of the resource being valued; employ an objective, statistically verifiable basis for scaling the results; and explain any divergence in values given for WTP and WTA by income or other effects consistent with standard consumer behavior.

[8] In the only case in which the issue was brought to trial, the court expressly disallowed the use of a CVM study to measure nonuse existence values of fish in a contaminated river. The court found that the study was "not persuasive" and it would have been "conjecture and speculation" to allow natural resource damages based on the study. *Idaho v. Southern Refrigerated Transport Company* [1991 WL 22479, p. 21 (D. Idaho 1991)]. See also *Mercado v. Ahmed* [756 F. Supp. 1097, 1103 (N.D. Ill. 1991)], aff'd [974 F.2d 863 (7th Cir. 1992)]. The court excluded an economist's proffered testimony as to value of a plaintiff's lost pleasure of life based on an amalgam of CVM willingness-to-pay studies; the court excluded testimony because of the variability and lack of economic consensus as to the validity and reliability of this application of contingent valuation methodology.

[9] Similar examples include the limits on liability that are imposed in wrongful death actions and recent state court refusals to award damages for the creation of a latent risk of incurring injury, such as a risk of heart valve failure or the risk of contracting a future illness as a result of exposure to a toxic substance. See, for example *Willett v. Baxter Int'l, Inc.* [929 F.2d 1094 (5th Cir. 1991)]; *Sill v. Shiley, Inc.* [No. 89-2077 WM (8th Cir. May 2, 1990)]; and *Hagepanos v. Shiley, Inc.* [846 F.2d 71 (4th Cir. 1988)].

[10] It may be, however, that regulations require more than the economically optimum degree of precaution and thereby impose costs on the regulated activity equal to or greater than the costs of residual injuries.

REFERENCES

Botkin, Daniel B. 1990. *Discordant Harmonies*. New York: Oxford University Press.

Cross, Frank B. 1993. Restoring Restoration for Natural Resource Damages. *University of Toledo Law Review* 24: 319, 334–35.

Cummings, R., and G. Harrison. 1992. *Identifying and Measuring Nonuse Values for Natural and Environmental Resources: A Critical Review of the State of the Art*. Unpublished working paper.

Desvousges, W., F. Johnson, R. Dunford, K. Boyle, S. Hudson, and K. Wilson. 1993. Measuring Natural Resources Damages with Contingent Valuation: Tests of Validity and Reliability. In *Contingent Valuation: A Critical Assessment*, edited by J. Hausman. Amsterdam: North Holland.

Dobbins, J.C. 1994. Note, The Pain and Suffering of Environmental Loss: Using Contingent Valuation to Estimate Nonuse Damages. *Duke Law Journal* 43: 879.

Duffield, J., and D. Patterson. 1992. *Field Testing Existence Value: An Instream Flow Trust Fund for Montana Rivers*. University of Montana. Unpublished working paper.

Freeman, A. M. 1993. Nonuse Values in Natural Resource Damage Assessment. In *Valuing Natural Assets*, edited by R.J. Kopp and V.K. Smith. Washington, D.C.: Resources for the Future.

Kahneman, D. 1986. Comments on the Contingent Valuation Method. In *Valuing Environmental Goods: An Assessment of the Contingent Valuation Method*, edited by R.G. Cummings and others. Savage, Maryland: Rowman and Littlefield.

Kahneman, D., and J.L. Knetsch. 1989. *Using Surveys to Value Public Goods*.

———. 1992a. Valuing Public Goods: The Purchase of Moral Satisfaction. *Journal of Environmental Economics and Management* 22: 57–70.

———. 1992b. Contingent Valuation and the Value of Public Goods: A Reply. *Journal of Environmental Economics and Management* 22: 90–94.

Kemp, M., and C. Maxwell. 1993. *Exploring a Budget Concept for Contingent Valuation Estimates*. In *Contingent Valuation: A Critical Assessment*, edited by J. Hausman. Amsterdam: North Holland.

Kopp, R.J., and V.K. Smith. 1993. *Valuing Natural Assets: The Economics of Natural Resource Damage Assessment*. Washington, D.C.: Resources for the Future.

McClelland, G.H., and others. 1992. *Methods for Measuring Non-Use Values: A Contingent Valuation Study of Groundwater Cleanup*. (October 1992). Washington, D.C.: U.S. EPA.

McFadden, D., and G. Leonard. 1993. Issues in the Contingent Valuation of Environmental Goods: Methodologies for Data Collection and Analysis. In *Contingent Valuation: A Critical Assessment*, edited by J. Hausman. Amsterdam: North Holland.

Mead, J. 1993. Review and Analysis of Recent State-of-the-Art Contingent Valuation Studies. In *Contingent Valuation: A Critical Assessment*, edited by J. Hausman. Amsterdam: North Holland.

Mitchell, R.C., and R.T. Carson. 1989. *Using Surveys to Value Public Goods: The Contingent Valuation Method.* Washington, D.C.: Resources for the Future.

NOAA (National Oceanic and Atmospheric Administration). 1993. Panel on Contingent Valuation. Report, 58 *Federal Register* 4601. Washington, D.C.: U.S. Government Printing Office.

Note, Ask a Silly Question. . . Contingent Valuation of Natural Resource Damages. 1992. *Harvard Law Review* 105: 1,981.

Rowe, R.D., and others. 1991. *Contingent Valuation of Natural Resource Damage Due to the Nestucca Oil Spill* (final report). Takoma, Washington.

Rubin, J., G. Helfand, and J. Loomis. 1992. A Benefit-Cost Analysis of the Northern Spotted Owl. *Journal of Forestry* 25–30.

Salmon Fishermen End Blockade of Alaska Port. 1993. *BNA Environment Reporter Current Developments* 781. Washington, D.C.: Bureau of Natural Affairs.

Schkade, D., and J. Payne. 1993. Where Do the Numbers Come From? How People Respond to Contingent Valuation Questions. In *Contingent Valuation: A Critical Assessment*, edited by J. Hausman. Amsterdam: North Holland.

Seip, K., and J. Strand. 1991. *Willingness to Pay for Environmental Goods in Norway: A Contingent Valuation Study With Real Payment.* Oslo: SAF Center For Applied Research, University of Oslo, Department of Economics.

Singer, P. 1975. *Animal Liberation.* New York: New York Review (Random House).

Smith, V.K. 1992. Arbitrary Values, Goods Causes, and Premature Verdicts: A Reaction to Kahneman and Knetsch. *Journal of Environmental Economics and Management* 22: 71–89.

Stevens, W. 1991. Balance of Nature? What Balance Is That? *New York Times* 22 October 1991, sec. C, p. 4.

Stewart, R. 1983. Regulation in a Liberal State: The Role of Non-Commodity Values. *Yale Law Journal* 92: 1537.

Taylor, P. 1986. *Respect for Nature.* Princeton: Princeton University Press.

U.S. GAO (General Accounting Office). 1993. *Natural Resources Restoration: Use of Exxon Valdez Oil Spill Settlement Funds.* GAO/RCED-93-206BR. Washington, D.C.: U.S. GAO.

APPENDIX

Conference Agenda and Participants

The following agenda outlines the conference "Superfund Reauthorization: Theoretical and Empirical Issues" held at the New York University School of Law in New York City on December 3–4, 1993. The subsequent list identifies the conference participants and their institutional affiliations.

Panel 1– Superfund Cleanups: Costs and Benefits

Moderator:
Richard D. Morgenstern

Papers:
W. Kip Viscusi and James T. Hamilton, "Human Health Risk Assessments for Superfund"
Katherine D. Walker, "Confronting Superfund Mythology: An Analysis of 1991 Records of Decision"
Maureen Cropper, "Do Benefits and Costs Matter in Environmental Regulation? An Analysis of EPA Decisions"

Commentators:
Robert W. Frantz, Peter Guerrero, David Case, Peter S. Menell

Panel 2 – The Problem of Transaction Costs

Moderator:
Edwin H. Clark

Papers:
Lloyd S. Dixon, "Superfund and Transaction Costs"
Lewis A. Kornhauser and Richard L. Revesz, "De Minimis Settlements Under Superfund"

Commentators:
Perry Beider, Leonard B. Barson

Panel 3 – Natural Resource Damages under Superfund

Moderator:
James E. Krier

Paper:
Richard B. Stewart, "Natural Resource Damages"

Commentators:
David F. Bradford, Gordon J. Johnson

Panel 4 – Alternatives to the Liability Scheme

Moderator:
Don R. Clay

Papers:
Katherine N. Probst, "The Economic Impact of Alternative Superfund Financing"
Lewis A. Kornhauser and Richard L. Revesz, "The Superfund Liability Scheme and Its Alternatives"

Commentators:
Kenneth S. Abraham, Edward Pollak, William J. Roberts, Peter S. Yu

Panel 5 – Focusing on the Reauthorization

Moderators:
Richard L. Revesz and Richard B. Stewart

CONFERENCE PARTICIPANTS AND AFFILIATIONS

Affiliations shown are those effective at the time of the conference.

Kenneth S. Abraham
Class of 1962 Professor of Law
University of Virginia School of Law

Leonard B. Barson
Counsel, Subcommittee on Transportation and Hazardous Materials, Committee on Energy and Commerce
U.S. House of Representatives

Perry Beider
Analyst, Natural Resources and Commerce Division
U.S. Congressional Budget Office

David F. Bradford
Professor of Economics and Public Affairs
Princeton University
Adjunct Professor of Law
New York University School of Law

APPENDIX 253

David Case
General Counsel
Hazardous Waste Treatment
Council

Edwin H. Clark
President
Clean Sites

Don R. Clay
President
Don Clay Associates, Inc.

Maureen Cropper
Principal Economist
World Bank
Professor of Economics
University of Maryland

Lloyd S. Dixon
Economist, Social Policy
Department
RAND Corporation

Robert W. Frantz
Manager and Counsel,
Environmental Remediation
Program
Corporate Environmental Programs
General Electric Company

Peter Guerrero
Associate Director, Environmental
Protection Issues
U.S. Government Accounting
Office

James T. Hamilton
Assistant Professor of Public
Health, Economics, and Political
Science
Duke University

Gordon J. Johnson
Deputy Bureau Chief,
Environmental Protection Bureau
New York State Department of Law

Lewis A. Kornhauser
Professor of Law
New York University School of Law

James E. Krier
Earl Warren DeLano Professor of Law
University of Michigan Law School

Peter S. Menell
Acting Professor of Law
University of California–Berkeley
School of Law

Richard D. Morgenstern
Director, Office of Policy Analysis
Office of Policy, Planning, and
Evaluation
U.S. Environmental Protection
Agency

Edward Pollak
Corporate Senior Vice President
Olin Corporation

Katherine N. Probst
Fellow, Center for Risk Management
Resources for the Future

Richard L. Revesz
Professor of Law
New York University School of Law

William J. Roberts
Legislative Director
Environmental Defense Fund

Richard B. Stewart
Professor of Law
New York University School of Law

W. Kip Viscusi
George G. Allen Professor of
Economics
Duke University

Katherine D. Walker
Research Associate
Harvard Center for Risk Analysis

Peter M. Yu
Director
National Economic Council

INDEX

Index

Air as exposure medium, 67
Alternative liability regimes, 17, 151–54
  cleanup costs, 146–51
  criteria for, 117–18
  industrial sectors effects, 159–65
  joint-and-several vs. nonjoint liability, 116–43
  taxes, 154–56
  trust fund in, 157–59
Alternative taxing regimes, 154–56
Amendments to Superfund. *See* SARA
ARARs (applicable or relevant and appropriate requirements), 13
  categories, 60
  risk assessment and, 44–48
  TBC goals and, 44, 52–53n.14
ASAP (Alliance for a Superfund Action Partnership), 151
Asbestos litigation, 182–83
Automobile accident litigation, 182–83

Banking industry and liability, 8–9
Baseline risk assessments, 26, 58–64
  See also Risk assessment and management
Benefit-cost analysis. *See* Cost-benefits of cleanups
Brookings Institution, 146
Business impacts. *See* Cleanup costs; Insurance industry; Private sector impacts

CAA (Clean Air Act), 60
Cancer risks, 26–27
  defined, 51n.3, 59
  extrapolation problems, 33
  maximum, 30–32, 34, 76–78
  pathways and risk levels, 62, 66, 70–78
Causation requirements, 7
CBO/GAO/RFF Survey, 185n.10
CERCLA (Comprehensive Environmental Response, Compensation, and Liability Act) statute, 4–13
  See also Reauthorization considerations
CERCLIS (CERCLA Information System), 10
Chemical and allied industries
  costs to, 17, 161–62, 163–64
  taxes on, 7–8, 154–56
Clean Air Act. *See* CAA
Clean Water Act. *See* CWA
Cleanup costs
  alternative regimes, 157–65
  cleanup standards and, 14–16
  cost-benefits, 15–16, 84–85
  funding for, 8
  future, 151
  industrial sectors and, 159–65
  liability and, 6, 117–18
  monitoring and, 118
  permanence of cleanups and, 100–1, 104–7
  recovering, 145–46, 157–65
  relative size, 165
  site types and, 148–49
  total cost estimates, 83, 109n.1, 165
  transaction costs and, 162–65, 177
  uncertainties, 145–46

257

258   INDEX

Cleanup of sites
  See also Cleanup costs; NPL sites
  decisions analyzed, 83–104
  liability and deterrence, 11, 116–28
  liability rules and, 117–18
  options, 94–104
  permanence of, 15, 95–96, 100–1, 104–7
  process leading to, 10–12
  relative size, 165
  transaction costs, 171–84
Cleanup stages, 11–12
  de minimis settlements and, 190–93, 208–9
Cleanup standards and goals, 12–16
  See also ARARs; Cancer risks; Noncancer risks
Cleanup technology options, 12–13
  cost-benefits, 94–104
Clinton administration proposal (H.R. 3800, 103rd Congress), 17, 116, 154
Comprehensive general liability, 9
Contingent valuation methodology. *See* CVM surveys
Cost-benefits of cleanups, 15–16, 83–104
Costs. *See* Cleanup costs; Cost-benefits of cleanups; Transaction costs
Creosote contamination. *See* Wood-preserving sites
CVM (contingent valuation methodology) surveys, 219–22
  commodification problems, 235, 238–39
  DOI use, 235
  NOAA panel assessment, 237
  noneconomic valuation by, 237–38
  nonuse values distortion, 234–38
CWA (Clean Water Act), 60
  NRD liability and, 219–20, 221, 226, 228

Damage assessment. *See* NRD liability

Databases
  CBO/GAO/RFF Survey, 185n.10
  CERCLIS (EPA), 10
  IRIS (EPA), 59
  NPL (EPA), 172–73, 185n.10
  NPL (RFF), 147
  RODs (HCRA), 28
  SSRS, 185n.10
De micromis exemptions, 151
De minimis settlements, 19
  cutoff criteria, 195–98, 213–14n.8
  EPA role, 187–89, 203–12
  instruments, 193–4
  number of, 188–89
  opposition, 211
  premiums, 201–4, 210
  pure vs. nonpure, 190–91
  reopeners criteria, 198–201, 209–210
  terms of settlements, 194–203
  timing, 190–94, 208–9
DOE (U.S. Department of Energy) sites, 149
DOI (U.S. Department of the Interior) regulations on NRD, 221, 228, 237, 242
  See also OPA

EIRF (Environmental Insurance Resolution Fund), 151–52, 158–60
EIT (Environmental Income Tax), 154–56
Environmental equity/justice, 16, 85, 91–94, 105–7
Environmental Income Tax. *See* EIT
Environmental Insurance Resolution Fund. *See* EIRF
Environmental restoration options
  See also NRD liability
  funding, 242–43
  costs, 222, 224–26, 229–30
Environmental risks, 29–30, 34–40
  See also Groundwater contamination; Soil contamination
EPA (U.S. Environmental Protection Agency)
  cleanup cost containment, 118

INDEX   259

cleanup criteria, 84–85
cleanup decisions analyzed, 83–103
cleanup process, 10–12
de minimis settlements, 187–88, 204–12
NPL database, 172–73, 185n.10
plaintiff role of, 18
risk assessment and, 25–28, 47–50, 58–60, 64–65, 72–73, 78–80
Exposure assessment, 59
Exposure media and routes, 60–61, 67
tabulated data, 66, 68, 71, 73, 75, 77
Exposure pathways, 62
See also Cancer risks; Noncancer risks; Risk pathways
Exxon *Valdez* oil spill, 220, 229, 230

Fairness issues in liability, 138–42
Future risks, 56, 61, 65–69, 72–76, 79–80, 91, 92, 93, 110n.7

Groundwater contamination, 67
See also Exposure media and routes
ARARs and, 44–47, 50
risk assessment and, 34–40
SDWA standards and, 13
tabulated data, 66, 68, 71, 73, 75, 77

H.R. 3800 (103rd Congress), 17, 116, 154
Harvard Center for Risk Analysis. *See* HCRA
Hazard indexes and quotients (noncancer), 27, 52n.4, 59–60
See also Noncancer risks
contaminants and index values, 40–42
Hazard Ranking System. *See* HRS
Hazardous Substances Response Trust Fund. *See* Trust fund
HCRA (Harvard Center for Risk Analysis), 26
Health risks, 27–30, 55–62
See also Cancer risks; Noncancer risks

chemicals contributing to, 40–42, 62–63
overstating, 32–34
HRS (Hazard Ranking System), 11, 91, 102–3

Insurance industry
cleanup costs, 164
EIRF and, 151
liability and, 9–10
transaction costs, 160, 164, 173–74, 177–79
IRIS (Integrated Risk Information System; EPA), 59

Joint-and-several liability, 3
curtailments suggested, 115–16
deterrence effects, 116–28
equilibria, 125–28
fairness issues, 138–42
H.R. 3800 changes, 116
nonjoint liability compared, 116–43
right of contribution and, 7
settlement incentives, 128–38

Land use risks, 39–40
future uses, 56, 61, 65–69, 72–76, 79–80
Legal costs. *See* Transaction costs
Legal profession and liability, 10
Liability regime, 3, 6–7, 152–53
See also Alternative liability regimes; De minimis settlements; Joint-and-several liability; NRD liability; Tort liability and litigation
alternatives, 16–17, 115–43, 145–65
criteria, 117–18
deterrence effects of, 116–28
H.R. 3800 changes, 116, 154
impacts on U.S. economy, 8–10
settlements and, 128–38, 140–41
solvency issues, 120–23, 136–37
LOAEL (Lowest Observed Adverse Effect Level), 51n.4, 60
Love Canal, 4–5

MCLGs (Maximum Contaminant Levels Goals), 13, 44–46
MCLs (Maximum Contaminant Levels), 13, 44–46
Mortgages and liability, 8–9
Municipalities liability, 10, 174

NAACP (National Association for the Advancement of Colored People), 151
National Contingency Plan. *See* NCP
National Oil and Hazardous Substances Pollution Contingency Plan, 74
National Priorities List. *See* NPL sites
Natural resource damages. *See* NRD liability
NCP (National Contingency Plan), 26, 46
NOAA (National Oceanic and Atmospheric Administration) regulations on NRD, 221, 228, 237, 24
  See also OPA
NOAEL (No Observed Adverse Effect Level), 51n.4, 60
Noncancer risks, 27, 59–60
  assessments, 51n.4
  maximum, 31–32, 34, 76–78
  pathways and risk levels, 62, 66, 70, 77
Nonjoint liability, 116–43
Nonuse valuation. *See* CVM surveys
NPL (National Priorities List) sites, 11–12
  cost characteristics, 148–51
  database, 147
  non–NPL site cleanups, 117–18
  studies, 25–47, 55–80, 83–107
NRD (natural resource damages) liability, 7, 20–21
  See also CVM surveys
  alternatives, 239
  cost potential, 158, 160, 164
  CVM role, 219–20

damage and injury assessment, 222–24
damages schedule system, 241–42
full compensation, 238–39
as hybrid system, 220–21
legal and institutional problems, 226–30
restoration funding, 242–43
restoration options and costs, 222, 224–26, 229–30
tort liability problems, 220–22, 230–32, 238–39, 244

Occupational risks, 37–40, 53n.13, 61–62
  tabulated data, 66, 68, 71, 73, 75, 77, 79
Oil Spill Liability Trust Fund, 155
Oil spills, 74, 229, 230
  See also OPA
OPA (Oil Pollution Act) and NRD liability, 220, 221, 237, 240, 242
Operation and maintenance (O&M) costs, 166n.4
Orphan shares, 7, 150, 152–54, 167n.12

PAHs (polyaromatic hydrocarbons) contamination, 88
PA (Preliminary Assessment), 11
Pathways. *See* Risk pathways
PCBs (polychlorinated biphenyls), 110n.4
  study of sites, 83–107,
  Westinghouse site risks, 69–70
PCP (pentachlorophenol; creosote) contamination, 88
Petroleum tax, 154–56
Population risks, 42–43, 61–62, 68–69
  tabulated data, 66, 68, 71, 73, 75, 77, 79
Potentially responsible parties. *See* PRPs
Preliminary Assessment (PA), 11
Private sector impacts, 8–10, 19

INDEX 261

See also Insurance industry
  cleanup costs, 159–62
  transaction costs, 162–65, 173–81
Prospective liability, 3, 117–18
PRPs (potentially responsible parties), 6
  See also De minimis settlements
  liability regimes and, 6–7, 115–16
PRP-lead cleanups, 12, 90
  See also RPs
  relative cleanup costs, 118
  transaction costs, 171–84
Public health. *See* Health risks

RAND studies on transaction costs, 8–9, 162, 172–84, 188, 191–92
RCRA (Resource Conservation and Recovery Act), 5, 45–46, 60, 240
RD/RA (Remedial Design/Remedial Action), 11–12
Real estate industry and liability, 8
Reauthorization considerations, 3–4
  cleanup standards, 14–16
  liability regime, 16–17
  natural resource damages, 20–21
  risk assessment, 47–50, 78–80, 105–7
  transaction costs, 17–19, 183–84
Records of Decision. *See* RODs
Reference dose. *See* RfD
Remedial actions
  See also Cleanup of sites
  ARARs and, 60
  risk assessment and, 26–28, 47–50
  soil cleanup options, 95–99
Remedial Design/Remedial Action. *See* RD/RA
Remedial Investigation/Feasibility Study. *See* RI/FS
Residential land use. *See* Land use risks
Resource Conservation and Recovery Act. *See* RCRA
Responsible parties. *See* RPs
Restoration options. *See* Environmental restoration options

Retroactive liability, 3, 6
Retrospective liability, 118
RfD (reference dose; noncancer), 27, 51n.4, 59–60
RI/FS (Remedial Investigation/Feasibility Study), 11
  risk assessment and, 59–60, 62–63, 90, 94, 105
Risk assessment and management, 26, 56
  See also Cancer risks; Environmental risks; Health risks; Noncancer risks; Risk pathways
  ARARs and, 44–48
  baseline risk assessments, 26, 58–64
  cleanup standards and, 14–15
  numerical criteria, 26-27, 51n.3–4, 59–60, 89–91
  pathway analysis, 60–62
  risk levels and pathways, 69–78
  site studies (RI/FS) in, 59–60, 62–63, 90, 94, 105
  studies, 25–50, 55–80, 84–107
  target risk levels, 84, 87, 89–94
Risk identification, 56–57
Risk levels and risk pathways, 69–78
Risk pathways, 60–62, 64–65
  See also Exposure media and routes
  assessment categories and, 65–68
  cancer risks, 70–78
  future, 56, 65–69, 72–76, 78–80
  maximum risks, 76–78
  noncancer risks, 76–78
  risk levels, 69–70
RME (reasonable maximum exposure), 63
RODs (Records of Decision), 11
  database (HCRA), 28
  risk summaries, 62–63
  studies using, 25–47, 55–80, 83–107, 188–89
RPM Survey, 147
RPs (responsible parties), 17, 145–46, 166n.5

See also PRPs
  cleanup costs, 148–49, 157–59, 163
  cleanups led by, 12, 149–51
  transaction costs, 147–50, 160, 162–63

Safe Drinking Water Act. *See* SDWA
SARA (Superfund Amendment and Reauthorization Act), 5
  cleanup requirements, 95–96, 100–1, 104–7
  liability under, 115
  risk assessment and, 26–27, 58
SDWA (Safe Drinking Water Act), 13, 46, 60, 165
Settlements
  See also De minimis settlements
  fairness and effects of, 140–41
  incentives and liability, 128–38
  litigation costs, 134–36
SI (Site Inspection), 11
Site cleanups. *See* Cleanup costs; Cleanup of sites; Cleanup standards and goals
Site Inspection. *See* SI
Site studies. *See* RI/FS
Small businesses, 19
Soil contamination
  ARARs and, 44–47, 50
  cleanup decisions, 84
  remedial alternatives, 95–98
  risk assessment and, 34–40
  study of PCB and PCP-PAH sites, 83–110
  tabulated data, 66, 68, 71, 73, 75, 77
Solvency and liability
  full, 120–23
  limited, 122–23, 136–37
SSRS Database, 185n.10
Strict liability, 3, 6, 119–22
Superfund statute, 4–13
  See also Reauthorization considerations
Superfund Amendment and Reauthorization Act. *See* SARA

Target risk levels, 84, 87, 89–94
Taxation regime, 7–8
  alternatives, 154–57
  EIRF, 151–52, 158–60
  EIT, 154–56
  NRD liability and, 242–43
TCB (to be considered) goals, 44, 52–53n.14
Third-party defenses, 6, 193–94
Tort liability and litigation
  NRD liability and, 220–22, 230–32, 238–39, 244
  transaction costs and, 182–83
Toxicity assessment, 59
Toxicity values, 40–42
Transaction-cost share, 174–75, 183–84
Transaction costs, 17–19, 171–72
  See also De minimis settlements
  alternative regimes, 160
  distribution among industries, 162–65
  firm characteristics and, 175–77, 184n.3
  government, 174, 181
  industry sector shares, 162–65
  insurers, 177–79
  interactions of players, 172–74
  private sector, 162–65, 173–81
  PRPs, 172–77, 179–81
  RPs, 147–50, 160, 162–63
  site characteristics and, 175–77, 184n.3–5
  size of, 165, 174–77
  tort litigation and, 182–83
  volumetric shares and, 175–76
Trust fund (Hazardous Substance Response Trust Fund), 17, 145–46, 157–60

University of Texas–Austin, 147

Valuation of natural resources
  See also CVM surveys
  commodification problems, 238–39
  nonuse values, 220, 232–39

passive use values, 226, 234–35
use values, 221, 230
VAT (value added tax), 156
Volumetric criteria in de minimis settlements, 195–200, 213–14n.8
Volumetric shares and costs, 175–76

Water contamination. *See* Exposure media and routes; Groundwater contamination
Water Pollution Control Act, 45
Welfare risks, 29, 50, 52n.12
Willingness to accept/pay. *See* WTA; WTP
Wood-preserving (creosote) sites
cleanup costs, 149
study, 83–107
WTA (Willingness to accept), 236, 245n.6–8
WTP (Willingness to pay), 100, 107, 236, 245n.6–8

Also of Interest from RFF

Assigning Liability for Superfund Cleanups:
An Analysis of Policy Options
Katherine N. Probst and Paul R. Portney

While more than 2,700 emergency removals of hazardous materials have taken place under Superfund, implementing the long-term cleanup program has been the object of considerable controversy. One of the most contentious issues is whether the liability standards in the law should be revised. The authors analyze the pros and cons associated with the current liability scheme and a variety of alternative liability approaches.

"Seems to be setting the agenda for reform of the liability standards under the 1980 Superfund statute."
—*World Insurance Report*

1992 • 62 pages • ISBN 0-915707-64-0 (paper)

Confronting Uncertainty in Risk Management:
A Guide for Decision-Makers
Adam M. Finkel

Providing a systematic way to think about, quantify, and respond to uncertainty in risk assessments, this report focuses on the ways in which uncertainty analysis can improve the quality of "routine" risk management actions.

1990 • 87 pages (paper)

Controlling Asbestos in Buildings:
An Economic Investigation
Donald N. Dewees

Concerns about the high exposure of workers during installation of asbestos in the past have been widely addressed. The problems posed by asbestos now present in existing buildings, however, are more difficult to deal with. The author develops a methodology for economic analysis of asbestos control programs in existing buildings and presents the results of three case studies.

1986 • 106 pages • ISBN 0-915707-27-6 (paper)

Also of Interest from RFF

**Economics and Episodic Disease:
The Benefits of Preventing a Giardiasis Outbreak**
Winston Harrington, Alan J. Krupnick, and Walter O. Spofford, Jr.
With exhaustive attention to detail, the authors estimate the social costs to a community arising from an outbreak of waterborne disease. Their appealing blend of economic theory and innovative empirical analysis will help to avoid contaminated drinking water and will enhance the study of food safety issues and public health episodes.
1991 • 202 pages (index) • ISBN 0-915707-59-4 (cloth)

**Footing the Bill for Superfund Cleanups:
Who Pays and How?**
*Katherine N. Probst, Don Fullerton, Robert E. Litan, and
Paul R. Portney*
The authors look at who pays the costs for cleaning up toxic waste sites under the current Superfund liability scheme on a site-by-site basis. They analyze the incidence of different taxing mechanisms and compare the financial effects on specific industries of the current Superfund program and of several alternative liability and tax mechanisms.
Copublished with the Brookings Institution
1995 • 176 pages • ISBN 0-8157-2994-4 (cloth) • ISBN 0-8157-2995-2 (paper)

**The Law and Policy of Toxic Substances Control:
A Case Study of Vinyl Chloride**
David D. Doniger
"A basic introduction to the rapidly evolving and increasingly important field of toxic substances."
—*Southern Economic Journal*
"Examines the complexity and fragmentation of the U.S. government programs to control toxic substances."
—*Journal of Economic Literature*
1978 • 179 pages • ISBN 0-8018-2235-1 (paper)

Also of Interest from RFF

Nuclear Imperatives and Public Trust:
Dealing with Radioactive Waste
Luther J. Carter

"Carter has done a masterful job of laying out the technical issues, the political maneuvering, and the governmental bungling that have occurred during the past three decades of the nuclear-power program."
—*Amicus Journal*

"Carter presents a detailed and penetrating analysis of the events and policy decisions that led to noncommunist countries' collective failure to manage their civilian nuclear waste problem...This is a valuable book that leaves the reader with a hopeful sense about the future...It is worthwhile reading for both newcomers and veterans of the nuclear debate."
—*Chemical and Engineering News*

"Refreshingly free of the partisanship that generally clouds the discussions of nuclear power."
—*The New York Times Book Review*

1987 • 473 pages (index) • ISBN 0-915707-47-0 (paper)

Readings in Risk
Theodore S. Glickman and Michael Gough, eds.

"A very practical and realistic publication."
—*Chemical and Engineering News*

"Could form the basis for a course in risk analysis. Little mathematical background is required, and each paper is followed by a set of questions for discussion... an excellent text to teach from."
—*American Scientist*

"Compiles the seminal essays on risk issues...presented in a convenient, objective, simple, and stimulating manner...Its organization, selection of papers, and concise but provocative introductory essays make it an understandable and desirable resource for a nontechnical audience...Has its greatest value as a classroom tool."
—*Environmental Science and Technology*

1990 • 262 pages • ISBN 0-915707-55-1 (paper)

Also of Interest from RFF

Worst Things First? The Debate over Risk-Based National Environmental Priorities
Edited by Adam M. Finkel and Dominic Golding
This book presents findings from a forum convened to explore the controversy over EPA's risk-based approach for setting the nation's environmental priorities. Agreeing that alternative ways exist to target the nation's resources for environmental protection, participants differ sharply as to whether these varied approaches complement each other or would disrupt environmental policy-making.
1994 • 346 pages • ISBN 0-915707-74-8 (cloth)